Editor
Nancy Hoffman

Managing Editor
Karen Goldfluss, M.S. Ed.

Editor-in-Chief
Sharon Coan, M.S. Ed.

Cover Artist
Barb Lorseyedi

Art Manager
Kevin Barnes

Art Director
CJae Froshay

Imaging
Rosa C. See

Product Manager
Phil Garcia

Publisher
Mary D. Smith, M.S. Ed.

Practice Makes Perfect Fractions

GRADE 5

Author

Mary Rosenberg

Teacher Created Resources, Inc.
6421 Industry Way
Westminster, CA, 92683
www.teachercreated.com
ISBN: 978-0-7439-8615-1
©2004 Teacher Created Resources, Inc.
Reprinted, 2012
Made in U.S.A.

Table of Contents

Introduction 3

Practice 1: Numerators and Denominators 4

Practice 2: Comparing Fractions 5

Practice 3: Fraction Bars 6

Practice 4: Naming Fractions 7

Practice 5: Adding and Subtracting Like Fractions 8

Practice 6: Finding Common Denominators 9

Practice 7: More Practice with Common Denominators 10

Practice 8: Adding and Subtracting Unlike Fractions 11

Practice 9: More Practice with Unlike Fractions 12

Practice 10: Equivalent Fractions 13

Practice 11: Renaming Fractions 14

Practice 12: Circle Fractions 15

Practice 13: Mixed Fractions 16

Practice 14: Writing Mixed Fractions 17

Practice 15: Adding and Subtracting Mixed Fractions 18

Practice 16: More Practice with Mixed Fractions 19

Practice 17: Money "Cents" 20

Practice 18: Improper Fractions 21

Practice 19: Adding and Subtracting Improper Fractions 22

Practice 20: Adding and Subtracting Mixed and Improper Fractions 23

Practice 21: Plotting Fractions 24

Practice 22: Multiplying Fractions 25

Practice 23: More Practice Multiplying Fractions 26

Practice 24: Dividing Fractions 27

Practice 25: More Practice Dividing Fractions 28

Practice 26: Fraction Sets 29

Practice 27: More Practice with Fraction Sets 30

Practice 28: Which is More? 31

Practice 29: Which is Less? 32

Practice 30: Finding Decimal Equivalents 33

Practice 31: Changing Fractions to Decimals and Decimals to Percents 34

Practice 32: More Practice with Fractions, Decimals, and Percents 35

Practice 33: Benchmark of 1/2 36

Practice 34: Benchmarks of 1/3 and 1/4 37

Practice 35: Comparing Fractions and Decimals 38

Practice 36: Fraction Word Problems 39

Test Practice Pages 40

Answer Sheet 46

Answer Key 47

Introduction

The old adage "practice makes perfect" can really hold true for your child and his or her education. The more practice and exposure your child has with concepts being taught in school, the more success he or she is likely to find. For many parents, knowing how to help your children can be frustrating because the resources may not be readily available. As a parent it is also difficult to know where to focus your efforts so that the extra practice your child receives at home supports what he or she is learning in school.

This book has been designed to help parents and teachers reinforce basic skills with their children. *Practice Makes Perfect* reviews basic math skills for children in grade 5. The math focus is on fractions. While it would be impossible to include all concepts taught in grade 5 in this book, the following basic objectives are reinforced through practice exercises. These objectives support math standards established on a district, state, or national level. (Refer to the Table of Contents for the specific objectives of each practice page.)

- writing and ordering fractions
- comparing proper fractions
- comparing equivalent fractions
- adding and subtracting fractions with like and unlike denominators
- adding and subtracting mixed and improper fractions
- writing fractions in simplest form
- multiplying and dividing fractions
- representing money as fractions

There are 36 practice pages organized sequentially, so children can build their knowledge from more basic skills to higher-level math skills. (Note: Have children show all work where computation is necessary to solve a problem. For multiple-choice responses on practice pages, children can fill in the letter choice or circle the answer.) Following the practice page are six test practices. These provide children with multiple-choice test items to help prepare them for standardized tests administered in schools. To correct the test pages and the practice pages in this book, use the Test Practice answer key provided on pages 47 and 48.

How to Make the Most of This Book

Here are some useful ideas for optimizing the practice pages in this book:

- Set aside a specific place in your home to work on the practice pages. Keep it neat and tidy with materials on hand.
- Set up a certain time of day to work on the practice pages. This will establish consistency. An alternative is to look for times in your day or week that are less hectic and conducive to practicing skills.
- Keep all practice sessions with your child positive and constructive. If the mood becomes tense or you and your child are frustrated, set the book aside and look for another time to practice with your child.
- Help with instructions if necessary. If your child is having difficulty understanding what to do or how to get started, work through the first problem with him or her.
- Review the work your child has done. This serves as reinforcement and provides further practice.
- Allow your child to use whatever writing instruments he or she prefers. For example, colored pencils can add variety and pleasure to drill work.
- Pay attention to the areas in which your child has the most difficulty. Provide extra guidance and exercises in those areas. Allowing children to use drawings and manipulatives, such as coins, tiles, game markers, or flash cards, can help them grasp difficult concepts more easily.
- Look for ways to make real-life applications of the skills being reinforced.

Practice 1

A **fraction** is a number that names part of a whole thing. The **numerator** is the number on the top and tells how many parts are being referred to. The **denominator** is the bottom number and shows how many equal parts there are in all.

Directions: Write what fraction of each shape is shaded.

1.	2.	3.

4.	5.	6.	7.

8.	9.	10.	11.

12.	13.	14.	15.

Practice 2

1.

$\frac{3}{10}$ ◯ $\frac{11}{12}$

2.

$\frac{6}{12}$ ◯ $\frac{2}{3}$

3.

$\frac{8}{10}$ ◯ $\frac{2}{5}$

4.

$\frac{2}{7}$ ◯ $\frac{4}{9}$

5.

$\frac{1}{8}$ ◯ $\frac{3}{7}$

6.

$\frac{3}{8}$ ◯ $\frac{7}{8}$

7.

$\frac{8}{9}$ ◯ $\frac{7}{9}$

8.

$\frac{6}{10}$ ◯ $\frac{4}{8}$

9.

$\frac{6}{9}$ ◯ $\frac{10}{12}$

10.

$\frac{1}{6}$ ◯ $\frac{1}{3}$

11.

$\frac{4}{5}$ ◯ $\frac{2}{10}$

12.

$\frac{3}{9}$ ◯ $\frac{3}{12}$

Practice 3

Directions: Shade each shape below to show the fraction. Then circle the fraction in each pair that is the largest.

1. $\frac{3}{9}$

   $\frac{3}{12}$

2. $\frac{2}{12}$

   $\frac{9}{11}$

3. $\frac{4}{10}$

   $\frac{1}{5}$

4. $\frac{5}{6}$

   $\frac{5}{10}$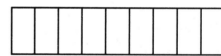

5. $\frac{5}{9}$

   $\frac{7}{10}$

6. $\frac{6}{8}$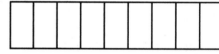

   $\frac{2}{9}$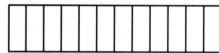

7. $\frac{5}{7}$

   $\frac{5}{11}$

8. $\frac{7}{9}$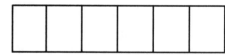

   $\frac{7}{12}$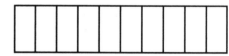

9. $\frac{1}{4}$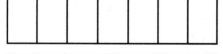

   $\frac{1}{12}$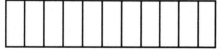

10. $\frac{3}{6}$

    $\frac{1}{10}$

11. $\frac{6}{11}$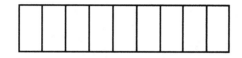

    $\frac{6}{7}$

12. $\frac{3}{7}$

    $\frac{2}{3}$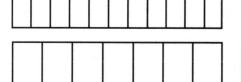

Practice 4

A **fraction** is a number that names part of a whole thing. The **numerator** is the number on the top and tells how many parts are being referred to. The **denominator** is the bottom number and shows how many equal parts there are in all.

Directions: Look at each ruler below, and write the fraction shown by the bold line. The first one has been done for you.

1.
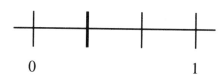
0   1

2.

0   1

3.

0   1

4.
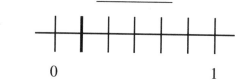
0   1

5.
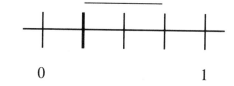
0   1

6.
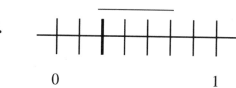
0   1

7.
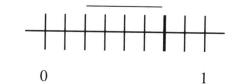
0   1

8.

0   1

9.
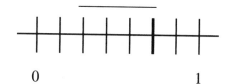
0   1

10.

0   1

11.
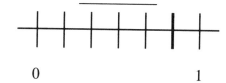
0   1

12.
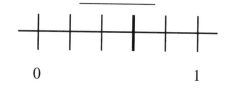
0   1

13.
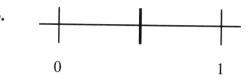
0   1

14.
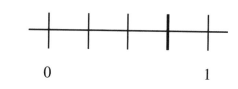
0   1

Practice 5

Fractions that have the same denominator are called **like fractions**. When adding or subtracting like fractions, the denominator does not change. For example: $\frac{1}{6} + \frac{4}{6} = \frac{5}{6}$ and $\frac{7}{8} - \frac{2}{8} = \frac{5}{8}$

- To add like fractions, add the numerators and write the sum over the denominator.
- To subtract like fractions, subtract one numerator from the other and write the difference over the denominator.

Directions: Solve the following problems by adding or subtracting as shown.

1. $\frac{1}{6} + \frac{3}{6} =$ _____

2. $\frac{7}{10} - \frac{4}{10} =$ _____

3. $\frac{10}{11} - \frac{1}{11} =$ _____

4. $\frac{4}{8} + \frac{3}{8} =$ _____

5. $\frac{1}{9} + \frac{3}{9} =$ _____

6. $\frac{3}{11} + \frac{5}{11} =$ _____

7. $\frac{3}{8} + \frac{1}{8} =$ _____

8. $\frac{4}{7} - \frac{3}{7} =$ _____

9. $\frac{6}{9} - \frac{5}{9} =$ _____

10. $\frac{1}{4} + \frac{2}{4} =$ _____

11. $\frac{6}{8} + \frac{1}{8} =$ _____

12. $\frac{2}{7} + \frac{4}{7} =$ _____

13. $\frac{3}{5} + \frac{1}{5} =$ _____

14. $\frac{9}{10} - \frac{6}{10} =$ _____

15. $\frac{4}{12} + \frac{7}{12} =$ _____

16. $\frac{1}{3} + \frac{1}{3} =$ _____

17. $\frac{2}{6} + \frac{2}{6} =$ _____

18. $\frac{2}{5} - \frac{1}{5} =$ _____

19. $\frac{8}{10} - \frac{5}{10} =$ _____

20. $\frac{11}{12} - \frac{5}{12} =$ _____

21. $\frac{4}{9} - \frac{2}{9} =$ _____

22. $\frac{7}{12} - \frac{2}{12} =$ _____

23. $\frac{7}{11} - \frac{2}{11} =$ _____

24. $\frac{2}{3} - \frac{1}{3} =$ _____

25. $\frac{3}{9}$ $+\frac{2}{9}$

26. $\frac{11}{12}$ $-\frac{2}{12}$

27. $\frac{1}{11}$ $+\frac{4}{11}$

28. $\frac{9}{11}$ $-\frac{8}{11}$

29. $\frac{1}{8}$ $+\frac{2}{8}$

30. $\frac{1}{5}$ $+\frac{1}{5}$

31. $\frac{6}{9}$ $+\frac{1}{9}$

32. $\frac{5}{8}$ $-\frac{1}{8}$

33. $\frac{3}{10}$ $+\frac{6}{10}$

34. $\frac{7}{11}$ $-\frac{6}{11}$

35. $\frac{3}{12}$ $+\frac{6}{12}$

36. $\frac{1}{6}$ $+\frac{3}{6}$

37. $\frac{4}{5}$ $-\frac{1}{5}$

38. $\frac{1}{12}$ $+\frac{6}{12}$

39. $\frac{4}{9}$ $-\frac{2}{9}$

40. $\frac{2}{10}$ $+\frac{3}{10}$

41. $\frac{9}{12}$ $-\frac{5}{12}$

42. $\frac{4}{10}$ $+\frac{4}{10}$

43. $\frac{8}{10}$ $-\frac{1}{10}$

44. $\frac{2}{8}$ $+\frac{4}{8}$

#8615 Practice Makes Perfect: Fractions

Practice 6 ꩜ ꩜ ꩜ ꩜ ꩜ ꩜ ꩜ ꩜ ꩜ ꩜ ꩜ ꩜

Directions: Write the **multiples** for each pair of numbers below. For example, the multiples of 3 are 3, 6, 9, 12, and so on (these are the products of 3 x 1, 3 x 2, 3 x 3, 3 x 4, etc.). Then circle the numbers in each pair that are the same. These are **common denominators**.

1. 5: _____, _____, _____, _____, _____, _____, _____, _____, _____, _____

10: _____, _____, _____, _____, _____, _____, _____, _____, _____, _____

2. 2: _____, _____, _____, _____, _____, _____, _____, _____, _____, _____

10: _____, _____, _____, _____, _____, _____, _____, _____, _____, _____

3. 3: _____, _____, _____, _____, _____, _____, _____, _____, _____, _____

9: _____, _____, _____, _____, _____, _____, _____, _____, _____, _____

4. 3: _____, _____, _____, _____, _____, _____, _____, _____, _____, _____

4: _____, _____, _____, _____, _____, _____, _____, _____, _____, _____

5. 5: _____, _____, _____, _____, _____, _____, _____, _____, _____, _____

7: _____, _____, _____, _____, _____, _____, _____, _____, _____, _____

6. 4: _____, _____, _____, _____, _____, _____, _____, _____, _____, _____

6: _____, _____, _____, _____, _____, _____, _____, _____, _____, _____

7. 6: _____, _____, _____, _____, _____, _____, _____, _____, _____, _____

7: _____, _____, _____, _____, _____, _____, _____, _____, _____, _____

8. 2: _____, _____, _____, _____, _____, _____, _____, _____, _____, _____

8: _____, _____, _____, _____, _____, _____, _____, _____, _____, _____

9. 8: _____, _____, _____, _____, _____, _____, _____, _____, _____, _____

9: _____, _____, _____, _____, _____, _____, _____, _____, _____, _____

Practice 7

Directions: Find the smallest common denominator for each pair of fractions and rewrite the fractions using the new denominator. Remember to multiply both the numerator and the denominator by the same number. Then solve the problem. The first one has been done for you.

Problem	Smallest Common Denominator	Multiply First Fraction by...	Multiply Second Fraction by...	Rewrite and Solve the Problem
1. $\frac{1}{4}+\frac{3}{7}$	28	7: $\frac{1}{4} \times \frac{7}{7} = \frac{7}{28}$	4: $\frac{3}{7} \times \frac{4}{4} = \frac{12}{28}$	$\frac{7}{28} + \frac{12}{28} = \frac{19}{28}$
2. $\frac{5}{6}-\frac{5}{8}$				
3. $\frac{2}{3}-\frac{6}{9}$				
4. $\frac{6}{7}-\frac{2}{10}$				
5. $\frac{2}{3}-\frac{1}{10}$				
6. $\frac{7}{10}-\frac{6}{9}$				
7. $\frac{6}{8}-\frac{5}{9}$				
8. $\frac{4}{9}-\frac{2}{7}$				
9. $\frac{3}{7}-\frac{1}{3}$				
10. $\frac{2}{3}-\frac{2}{5}$				
11. $\frac{3}{10}+\frac{3}{5}$				
12. $\frac{5}{6}-\frac{3}{7}$				
13. $\frac{1}{3}-\frac{1}{4}$				

 #8615 Practice Makes Perfect: Fractions

Practice 8

Directions: Find the common denominator and rewrite the problem using the new denominator. Remember to multiply both the numerator and the denominator by the same number. Then add or subtract to solve each problem. Reduce the answer to its simplest form.

1. (A) $\dfrac{2}{12} - \dfrac{1}{9}$

2. (C) $\dfrac{2}{7} + \dfrac{2}{4}$

3. (E) $\dfrac{3}{4} - \dfrac{6}{10}$

4. (F) $\dfrac{2}{5} + \dfrac{6}{12}$

5. (I) $\dfrac{5}{6} - \dfrac{8}{10}$

6. (L) $\dfrac{3}{6} - \dfrac{2}{11}$

7. (M) $\dfrac{3}{9} - \dfrac{1}{10}$

8. (N) $\dfrac{1}{5} + \dfrac{1}{6}$

9. (O) $\dfrac{4}{5} - \dfrac{4}{6}$

10. (R) $\dfrac{9}{10} - \dfrac{3}{5}$

11. (S) $\dfrac{8}{9} - \dfrac{1}{12}$

12. (T) $\dfrac{8}{12} - \dfrac{1}{3}$

13. (Z) $\dfrac{7}{12} - \dfrac{5}{9}$

Directions: Write the letter from each answer above on the lines to decode the hidden message.

—————   —————   —————   —————   —————   —————   —————   —————   —————
9/10   3/10   1/18   11/14   1/3   1/30   2/15   11/30   29/36

—————   —————   —————   —————   —————   —————   —————   —————   —————
11/14   2/15   7/30   3/20   1/30   11/30   1/18   7/22   7/22

          —————   —————   —————   —————   —————
          29/36   1/30   1/36   3/20   29/36 .

Practice 9

Directions: Find the common denominator and rewrite the problem using the new denominator. Remember to multiply both the numerator and the denominator by the same number. Then add or subtract to solve each problem. Reduce the answer to its simplest form.

1. $\frac{10}{12}$ $-\frac{1}{5}$

2. $\frac{3}{10}$ $+\frac{1}{3}$

3. $\frac{9}{10}$ $-\frac{6}{7}$

4. $\frac{1}{8}$ $+\frac{8}{10}$

5. $\frac{6}{9}$ $-\frac{1}{2}$

6. $\frac{5}{6}$ $-\frac{6}{10}$

7. $\frac{7}{8}$ $-\frac{2}{6}$

8. $\frac{8}{9}$ $-\frac{3}{11}$

9. $\frac{3}{9}$ $+\frac{3}{8}$

10. $\frac{6}{8}$ $-\frac{7}{10}$

11. $\frac{2}{4}$ $-\frac{4}{8}$

12. $\frac{1}{10}$ $+\frac{4}{5}$

13. $\frac{4}{6}$ $-\frac{2}{5}$

14. $\frac{5}{9}$ $-\frac{4}{12}$

15. $\frac{2}{8}$ $+\frac{1}{7}$

Directions: Use the > (greater than), < (less than), or = (equal to) symbols to compare each set of fractions.

16. $\frac{2}{3}$ ◯ $\frac{11}{12}$

17. $\frac{4}{6}$ ◯ $\frac{2}{5}$

18. $\frac{1}{6}$ ◯ $\frac{1}{4}$

19. $\frac{4}{7}$ ◯ $\frac{2}{9}$

20. $\frac{3}{4}$ ◯ $\frac{1}{9}$

21. $\frac{3}{12}$ ◯ $\frac{3}{7}$

22. $\frac{5}{10}$ ◯ $\frac{7}{12}$

23. $\frac{1}{12}$ ◯ $\frac{3}{5}$

24. $\frac{4}{9}$ ◯ $\frac{8}{12}$

25. $\frac{6}{12}$ ◯ $\frac{5}{9}$

26. $\frac{4}{12}$ ◯ $\frac{4}{10}$

27. $\frac{2}{10}$ ◯ $\frac{2}{8}$

28. $\frac{5}{12}$ ◯ $\frac{1}{7}$

29. $\frac{2}{7}$ ◯ $\frac{5}{8}$

30. $\frac{2}{12}$ ◯ $\frac{7}{9}$

Practice 10 ❧ ❧ ❧ ❧ ❧ ❧ ❧ ❧ ❧ ❧ ❧ ❧ ❧

Equivalent fractions are fractions that name the same amount, such as 1/2 and 2/4. To tell if fractions are equivalent, reduce each fraction to its simplest form by dividing both the numerator and denominator by the same, largest possible number.

For example: 2/3 and 6/9

$6/9 \div 3/3 = 2/3$ so 2/3 and 6/9 are equivalent fractions

Directions: Circle the fractions in each row that are equivalent to the fraction in the first column. The first one has been done for you.

1.	$\frac{3}{12}$	$\boxed{\frac{1}{4}}$	$\frac{8}{11}$	$\frac{2}{8}$	$\frac{3}{10}$
2.	$\frac{1}{8}$	$\frac{4}{8}$	$\frac{3}{24}$	$\frac{4}{12}$	$\frac{2}{16}$
3.	$\frac{5}{6}$	$\frac{5}{8}$	$\frac{25}{30}$	$\frac{5}{7}$	$\frac{10}{12}$
4.	$\frac{1}{4}$	$\frac{4}{16}$	$\frac{10}{11}$	$\frac{9}{11}$	$\frac{3}{12}$
5.	$\frac{3}{5}$	$\frac{4}{5}$	$\frac{6}{10}$	$\frac{11}{12}$	$\frac{15}{25}$
6.	$\frac{2}{7}$	$\frac{8}{28}$	$\frac{5}{12}$	$\frac{5}{9}$	$\frac{6}{21}$
7.	$\frac{2}{3}$	$\frac{3}{8}$	$\frac{6}{9}$	$\frac{8}{12}$	$\frac{8}{10}$
8.	$\frac{3}{4}$	$\frac{7}{10}$	$\frac{9}{12}$	$\frac{6}{7}$	$\frac{18}{24}$
9.	$\frac{1}{6}$	$\frac{5}{26}$	$\frac{1}{5}$	$\frac{5}{30}$	$\frac{3}{18}$
10.	$\frac{1}{4}$	$\frac{2}{11}$	$\frac{5}{20}$	$\frac{2}{8}$	$\frac{1}{12}$
11.	$\frac{2}{9}$	$\frac{4}{18}$	$\frac{4}{28}$	$\frac{6}{27}$	$\frac{2}{3}$
12.	$\frac{1}{3}$	$\frac{6}{11}$	$\frac{2}{6}$	$\frac{3}{9}$	$\frac{3}{6}$
13.	$\frac{3}{5}$	$\frac{15}{25}$	$\frac{7}{9}$	$\frac{6}{9}$	$\frac{6}{10}$
14.	$\frac{1}{3}$	$\frac{1}{9}$	$\frac{4}{12}$	$\frac{3}{12}$	$\frac{6}{18}$
15.	$\frac{1}{2}$	$\frac{5}{10}$	$\frac{4}{11}$	$\frac{2}{10}$	$\frac{50}{100}$

Renaming Fractions

Practice 11

Directions: Write the fraction shown on each number line in bold. Then write an equivalent fraction by multiplying both the numerator and denominator by the same number. The first one has been done for you.

1.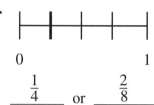

0   1

$\frac{1}{4}$ or $\frac{2}{8}$

2.

0   1

_____ or _____

3.

0   1

_____ or _____

4.

0   1

_____ or _____

5.

0   1

_____ or _____

6.

0   1

_____ or _____

7.

0   1

_____ or _____

8.

0   1

_____ or _____

9.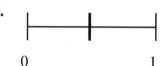

0   1

_____ or _____

10.

0   1

_____ or _____

11.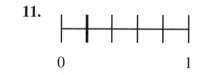

0   1

_____ or _____

12.

0   1

_____ or _____

Directions: Reduce each fraction below to its simplest form. For example: $\frac{6}{12} = \frac{1}{2}$

13. $\frac{4}{6}$ = _____

14. $\frac{8}{10}$ = _____

15. $\frac{9}{12}$ = _____

16. $\frac{6}{9}$ = _____

17. $\frac{2}{12}$ = _____

18. $\frac{3}{9}$ = _____

19. $\frac{4}{10}$ = _____

20. $\frac{6}{8}$ = _____

21. $\frac{3}{6}$ = _____

22. $\frac{10}{12}$ = _____

23. $\frac{2}{10}$ = _____

24. $\frac{3}{12}$ = _____

Practice 12

Directions: Write the fraction for the unmarked section in each circle below. Reduce each fraction to its simplest form.

1.

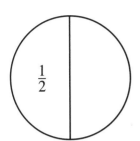

2.

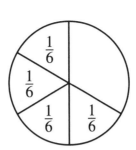

3.

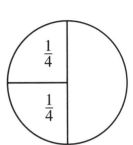

4.

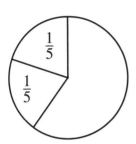

5.

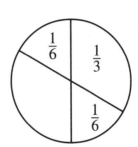

6.

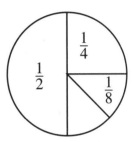

7.

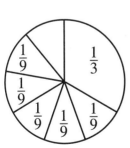

8.

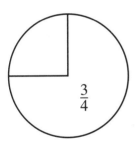

9.

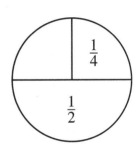

10.

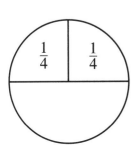

11.

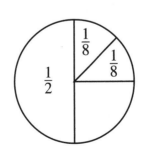

12.

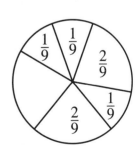

13.

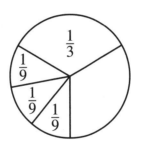

14.

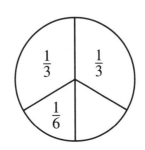

15.

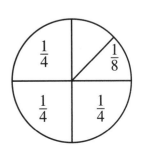

Practice 13

Directions: A **mixed fraction** contains both a whole number and a fraction. Write the mixed fraction for the shaded areas below. Reduce each fraction to its simplest form.

1.
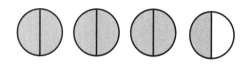

_____

2.
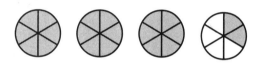

_____

3.
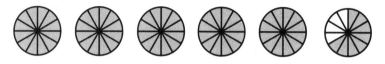

_____

4.
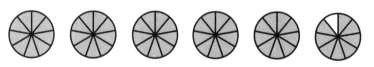

_____

5.

_____

6.
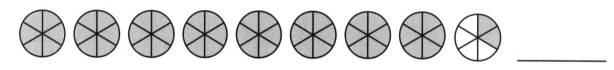

_____

7.
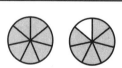

_____

8.

_____

9.
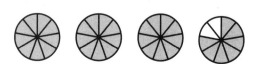

_____

10.
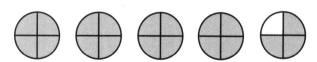

_____

Writing Mixed Fractions

Practice 14

Directions: A **mixed fraction** contains both a whole number and a fraction. Write the mixed fraction shown in bold on each number line.

1.

2.

3.

4.
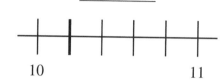

5.

6.

7.

8.
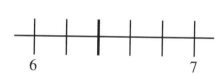

9.

10.

11.

12.
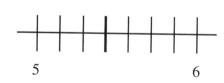

13.

14.
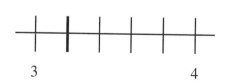

Practice 15 ꙮ ꙮ ꙮ ꙮ ꙮ ꙮ ꙮ ꙮ ꙮ ꙮ ꙮ ꙮ ꙮ

Directions: Add or subtract the fractions and reduce the answer to its simplest form.

1. $\quad 1\frac{6}{8}$
$+\ 9\frac{5}{8}$

2. $\quad 8\frac{8}{9}$
$-\ 3\frac{2}{9}$

3. $\quad 3\frac{4}{7}$
$+\ 7\frac{1}{7}$

4. $\quad 6\frac{9}{10}$
$-\ 2\frac{4}{10}$

5. $\quad 9\frac{7}{12}$
$-\ 5\frac{7}{12}$

6. $\quad 9\frac{8}{10}$
$-\ 4\frac{7}{10}$

7. $\quad 4\frac{5}{10}$
$-\ 3\frac{2}{10}$

8. $\quad 8\frac{3}{7}$
$+\ 3\frac{6}{7}$

9. $\quad 10\frac{5}{6}$
$-\ 5\frac{1}{6}$

10. $\quad 6\frac{6}{8}$
$+\ 7\frac{6}{8}$

11. $\quad 10\frac{2}{8}$
$+\ 7\frac{5}{8}$

12. $\quad 1\frac{4}{5}$
$-\ 1\frac{3}{5}$

13. $\quad 8\frac{1}{6}$
$+6\frac{2}{6}$

14. $\quad 3\frac{1}{2}$
$+\ 7\frac{1}{2}$

15. $\quad 8\frac{9}{12}$
$-\ 1\frac{8}{12}$

16. $\quad 5\frac{2}{5}$
$-\ 3\frac{2}{5}$

17. $\quad 1\frac{7}{11}$
$+\ 10\frac{7}{11}$

18. $\quad 9\frac{4}{12}$
$+\ 9\frac{7}{12}$

19. $\quad 4\frac{3}{10}$
$-\ 2\frac{3}{10}$

20. $\quad 10\frac{10}{12}$
$-\ 5\frac{1}{12}$

21. $\quad 4\frac{3}{4}$
$-\ 3\frac{1}{4}$

22. $\quad 2\frac{4}{6}$
$+\ 4\frac{4}{6}$

23. $\quad 9\frac{6}{7}$
$-\ 6\frac{4}{7}$

24. $\quad 6\frac{8}{10}$
$+\ 7\frac{9}{10}$

25. $\quad 9\frac{7}{8}$
$-\ 9\frac{4}{8}$

Practice 16

Directions: Add the fractions and reduce the answer to its simplest form.

1. $3\frac{1}{2}$
   $+\ 3\frac{1}{2}$

2. $3\frac{2}{4}$
   $+\ 7\frac{1}{4}$

3. $1\frac{1}{4}$
   $+\ 2\frac{3}{4}$

4. $5\frac{6}{7}$
   $+\ 1\frac{1}{7}$

5. $3\frac{4}{8}$
   $+\ 4\frac{4}{8}$

6. $8\frac{3}{9}$
   $+\ 2\frac{8}{9}$

7. $7\frac{5}{9}$
   $+\ 1\frac{1}{9}$

8. $1\frac{3}{5}$
   $+\ 6\frac{2}{5}$

9. $4\frac{6}{11}$
   $+\ 6\frac{5}{11}$

10. $1\frac{6}{12}$
    $+\ 1\frac{3}{12}$

11. $4\frac{4}{9}$
    $+\ 1\frac{6}{9}$

12. $4\frac{2}{8}$
    $+\ 1\frac{2}{8}$

13. $8\frac{3}{6}$
    $+\ 1\frac{1}{6}$

14. $3\frac{5}{10}$
    $+\ 7\frac{9}{10}$

15. $3\frac{7}{11}$
    $+\ 2\frac{5}{11}$

16. $2\frac{4}{12}$
    $+\ 8\frac{1}{12}$

17. $7\frac{10}{11}$
    $+\ 9\frac{7}{11}$

18. $3\frac{11}{12}$
    $+\ 9\frac{4}{12}$

19. $8\frac{3}{4}$
    $+\ 2\frac{1}{4}$

20. $8\frac{9}{11}$
    $+\ 9\frac{2}{11}$

21. $2\frac{7}{8}$
    $+\ 1\frac{5}{8}$

22. $6\frac{4}{5}$
    $+\ 1\frac{4}{5}$

23. $1\frac{4}{11}$
    $+\ 2\frac{6}{11}$

24. $1\frac{6}{12}$
    $+\ 3\frac{4}{12}$

Practice 17

| A dollar is worth 100 cents. Therefore, one cent is 1/100 of a dollar. | |

Directions: Write each amount in standard form.

1. $ $\frac{8}{100}$ _____

2. $10 $\frac{82}{100}$ _____

3. $64 $\frac{78}{100}$ _____

4. $ $\frac{32}{100}$ _____

5. $3 $\frac{66}{100}$ _____

6. $44 $\frac{50}{100}$ _____

7. $ $\frac{17}{100}$ _____

8. $9 $\frac{49}{100}$ _____

9. $93 $\frac{14}{100}$ _____

10. $3 $\frac{81}{100}$ _____

11. $34 $\frac{15}{100}$ _____

12. $51 $\frac{72}{100}$ _____

Directions: Write each amount as a fraction or a mixed fraction.

13. $.88

14. $3.12

15. $31.02

16. $.55

17. $9.27

18. $81.89

19. $.72

20. $5.32

21. $75.65

22. $.02

23. $6.03

24. $23.57

Directions: Solve each problem.

25. $.97
      + $.58

26. $1.23
      − $.87

27. $.85
      + $.62

28. $.73
      − $.49

29. $3.80
      + $8.10

30. $1.66
      + $9.34

31. $10.49
      − $4.78

32. $4.68
      + $5.13

33. $24.12
      + $34.24

34. $36.85
      + $64.11

35. $19.57
      + $89.62

36. $23.35
      + $25.19

37. $10.10
      + $34.79

38. $50.70
      − $10.74

39. $36.88
      − $15.19

Practice 18

An **improper fraction** is a fraction that has a numerator larger than or equal to its denominator.

Directions: Circle the improper fractions in each of the following rows.

1. $\frac{33}{10}$ $\frac{1}{4}$ $\frac{5}{6}$ $\frac{7}{3}$ 9 $\frac{35}{7}$

2. $\frac{17}{4}$ $\frac{5}{6}$ $\frac{41}{10}$ $\frac{3}{5}$ $\frac{12}{5}$ $\frac{4}{5}$

3. $\frac{13}{2}$ $\frac{1}{2}$ $\frac{6}{9}$ $\frac{10}{9}$ $\frac{12}{11}$ $\frac{1}{7}$

4. $\frac{20}{5}$ 3 $\frac{1}{9}$ $\frac{50}{6}$ $\frac{2}{3}$ $\frac{8}{3}$

Directions: Change each improper fraction below into a mixed fraction. To do so, divide the numerator by the denominator, and then rewrite it as a mixed fraction having a whole number and a fraction. The fraction should be written in simplest form.

5. $\frac{14}{10}$ 6. $\frac{18}{4}$ 7. $\frac{17}{5}$

8. $\frac{33}{5}$ 9. $\frac{16}{4}$ 10. $\frac{50}{12}$

11. $\frac{27}{7}$ 12. $\frac{33}{5}$ 13. $\frac{70}{11}$

14. $\frac{14}{10}$ 15. $\frac{57}{12}$ 16. $\frac{119}{12}$

Directions: Change each mixed fraction below into an improper fraction. To do so, multiply the whole number by the denominator and add that sum to the numerator. Then write the sum over the denominator.

17. $6\frac{3}{8}$ 18. $1\frac{4}{5}$ 19. $8\frac{7}{10}$

20. $8\frac{3}{7}$ 21. $8\frac{1}{12}$ 22. $7\frac{8}{10}$

23. $1\frac{7}{12}$ 24. $2\frac{1}{3}$ 25. $1\frac{1}{4}$

26. $2\frac{5}{6}$ 27. $6\frac{2}{11}$ 28. $4\frac{2}{3}$

29. $2\frac{3}{12}$ 30. $2\frac{6}{10}$

Adding and Subtracting Improper Fractions

Practice 19 ෬ ෭ ෬ ෭ ෬ ෭ ෬ ෭ ෬ ෭ ෬ ෭ ෬ ෭ ෬

Directions: Solve each of the problems below. Reduce each answer to its simplest form.

1. $\frac{5}{6} + \frac{8}{6}$

2. $\frac{6}{3} - \frac{2}{3}$

3. $\frac{15}{12} + \frac{6}{12}$

4. $\frac{10}{5} - \frac{2}{5}$

5. $\frac{11}{9} + \frac{9}{9}$

6. $\frac{14}{9} - \frac{6}{9}$

7. $\frac{14}{11} + \frac{4}{11}$

8. $\frac{17}{11} - \frac{6}{11}$

9. $\frac{15}{12} - \frac{4}{12}$

10. $\frac{16}{11} + \frac{2}{11}$

11. $\frac{15}{12} + \frac{9}{12}$

12. $\frac{9}{8} + \frac{6}{8}$

13. $\frac{8}{7} - \frac{4}{7}$

14. $\frac{9}{6} - \frac{4}{6}$

15. $\frac{10}{12} + \frac{13}{12}$

16. $\frac{19}{12} + \frac{7}{12}$

17. $\frac{13}{9} + \frac{8}{9}$

18. $\frac{12}{10} + \frac{6}{10}$

19. $\frac{7}{4} - \frac{2}{4}$

20. $\frac{6}{5} - \frac{3}{5}$

21. $\frac{9}{5} + \frac{3}{5}$

22. $\frac{10}{6} - \frac{8}{6}$

23. $\frac{7}{3} + \frac{11}{3}$

24. $\frac{7}{5} - \frac{4}{5}$

25. $\frac{11}{9} + \frac{10}{9}$

26. $\frac{19}{9} - \frac{8}{9}$

27. $\frac{22}{11} - \frac{15}{11}$

28. $\frac{13}{9} - \frac{8}{9}$

29. $\frac{9}{8} + \frac{6}{8}$

30. $\frac{6}{4} + \frac{5}{4}$

Practice 20

Directions: For each problem, rewrite the mixed fractions as improper fractions. Then solve the problem. Reduce the answer to its simplest form.

1. $8\frac{5}{12} - \frac{25}{12}$

2. $6\frac{1}{4} - \frac{20}{4}$

3. $5\frac{10}{11} + \frac{37}{11}$

4. $1\frac{8}{10} + \frac{12}{10}$

5. $8\frac{3}{6} + \frac{40}{6}$

6. $3\frac{5}{10} + \frac{32}{10}$

7. $5\frac{2}{11} - \frac{44}{12}$

8. $3\frac{5}{10} - \frac{23}{10}$

9. $10\frac{4}{9} + \frac{53}{9}$

10. $7\frac{1}{8} - \frac{29}{8}$

11. $7\frac{7}{12} + \frac{47}{12}$

12. $1\frac{2}{9} - \frac{10}{9}$

13. $3\frac{1}{3} - \frac{8}{3}$

14. $1\frac{5}{12} - \frac{13}{12}$

15. $5\frac{5}{9} - \frac{36}{9}$

16. $1\frac{9}{12} + \frac{15}{12}$

17. $8\frac{1}{5} - \frac{37}{5}$

18. $2\frac{8}{10} - \frac{19}{10}$

19. $5\frac{2}{5} - \frac{22}{5}$

20. $3\frac{1}{10} - \frac{16}{10}$

Directions: Use the symbols > (greater than), < (less than), or = (equal to) to compare each pair of the fractions.

21. $9\frac{1}{7}$ ◯ $4\frac{8}{7}$

22. $6\frac{5}{7}$ ◯ $\frac{69}{7}$

23. $9\frac{6}{8}$ ◯ $\frac{33}{8}$

24. $10\frac{1}{9}$ ◯ $7\frac{3}{9}$

25. $7\frac{3}{7}$ ◯ $\frac{59}{7}$

26. $4\frac{2}{3}$ ◯ $\frac{26}{3}$

27. $2\frac{7}{8}$ ◯ $\frac{80}{8}$

28. $6\frac{5}{6}$ ◯ $\frac{27}{6}$

29. $3\frac{2}{6}$ ◯ $\frac{19}{6}$

30. $7\frac{2}{7}$ ◯ $\frac{36}{7}$

31. $5\frac{3}{4}$ ◯ $\frac{8}{4}$

32. $3\frac{6}{10}$ ◯ $\frac{93}{10}$

33. $7\frac{2}{5}$ ◯ $\frac{42}{5}$

34. $7\frac{5}{10}$ ◯ $\frac{13}{4}$

35. $9\frac{7}{10}$ ◯ $\frac{82}{10}$

Practice 21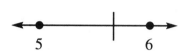

Directions: Circle the amount shown on each number line.

1.

8   9

$8\frac{1}{4}$ $8\frac{1}{2}$ $8\frac{3}{4}$

2.

4   5

$4\frac{1}{2}$ $4\frac{1}{4}$ $5\frac{1}{4}$

3.

5   6

$5\frac{1}{4}$ $5\frac{2}{3}$ $6\frac{2}{3}$

4.

7   8

$7\frac{1}{4}$ $7\frac{3}{4}$ $8\frac{1}{2}$

5.

3   4

$3\frac{1}{3}$ $3\frac{1}{2}$ $3\frac{3}{4}$

6.

9   10

$9\frac{1}{4}$ $9\frac{1}{2}$ $9\frac{2}{3}$

7.

6   7

$6\frac{1}{3}$ $6\frac{2}{3}$ $7\frac{2}{3}$

8.

2   3

$2\frac{1}{4}$ $2\frac{1}{2}$ $2\frac{3}{4}$

9.

1   2

$1\frac{1}{4}$ $1\frac{3}{4}$ $2\frac{1}{4}$

Directions: Place the following fractions on the number line.

0   1   2   3   4   5   6

10. $2\frac{1}{2}$ **11.** $5\frac{1}{3}$ **12.** $3\frac{2}{3}$ **13.** $4\frac{1}{4}$ **14.** $\frac{1}{2}$ **15.** $1\frac{3}{4}$

Directions: On the line next to each letter, write the fraction that each line segment represents.

0   *A*   *I*   1   *E*   2   *B*   *J*   3 *L F H*   4   *C*   5   *K G D*   6

A _____   D _____   G _____   J _____

B _____   E _____   H _____   K _____

C _____   F _____   I _____   L _____

#8615 Practice Makes Perfect: Fractions   *© Teacher Created Resources, Inc.*

Practice 22

To multiply fractions:

    1. Multiply the two numerators together.

    2. Multiply the two denominators together.

    3. Reduce the answer to its simplest form.

$$\frac{2}{3} \times \frac{3}{5} = \frac{6}{15} = \frac{2}{5}$$

Directions: Solve the problems, and use the letter for each answer to decode the hidden message at the bottom.

1. (A) $\frac{8}{12} \times \frac{1}{4} =$

2. (E) $\frac{1}{12} \times \frac{1}{5} =$

3. (H) $\frac{3}{4} \times \frac{5}{10} =$

4. (I) $\frac{3}{7} \times \frac{2}{5} =$

5. (N) $\frac{6}{12} \times \frac{1}{8} =$

6. (O) $\frac{9}{10} \times \frac{9}{12} =$

7. (R) $\frac{2}{9} \times \frac{7}{12} =$

8. (S) $\frac{2}{4} \times \frac{3}{8} =$

9. (T) $\frac{1}{7} \times \frac{4}{10} =$

10. (W) $\frac{1}{3} \times \frac{6}{9} =$

11. (Y) $\frac{2}{11} \times \frac{9}{11} =$

Hidden Message

___ ___ ___ ,   ___ ___   ___ ___ ___ ___
$\frac{6}{35}$ $\frac{2}{35}$ $\frac{3}{16}$ $\frac{1}{6}$ $\frac{3}{16}$ $\frac{1}{60}$ $\frac{1}{6}$ $\frac{3}{16}$ $\frac{18}{121}$

___ ___   ___ ___ ___ ,   ___ ___ ___ ,
$\frac{1}{6}$ $\frac{3}{16}$ $\frac{27}{40}$ $\frac{1}{16}$ $\frac{1}{60}$ $\frac{2}{35}$ $\frac{2}{9}$ $\frac{27}{40}$

   ___ ___ ___ ___ ___ !
$\frac{2}{35}$ $\frac{3}{8}$ $\frac{7}{54}$ $\frac{1}{60}$ $\frac{1}{60}$

Practice 23 ⟳ ☙ ⟳ ☙ ⟳ ☙ ⟳ ☙ ⟳ ☙ ⟳ ☙ ⟳ ⟳ ☙

To multiply fractions:

    1. Multiply the two numerators together.

    2. Multiply the two denominators together.

    3. Reduce the answer to its simplest form.

$$\frac{2}{3} \times \frac{3}{5} = \frac{6}{15} = \frac{2}{5}$$

Directions: Solve the problems below.

1. $\frac{9}{11} \times \frac{2}{3} =$ _____

2. $\frac{7}{9} \times \frac{4}{11} =$ _____

3. $\frac{3}{8} \times \frac{1}{2} =$ _____

4. $\frac{4}{6} \times \frac{3}{9} =$ _____

5. $\frac{3}{12} \times \frac{1}{3} =$ _____

6. $\frac{5}{7} \times \frac{3}{7} =$ _____

7. $\frac{1}{6} \times \frac{3}{5} =$ _____

8. $\frac{1}{4} \times \frac{2}{6} =$ _____

9. $\frac{1}{10} \times \frac{5}{9} =$ _____

10. $\frac{4}{8} \times \frac{1}{12} =$ _____

11. $\frac{6}{10} \times \frac{3}{11} =$ _____

12. $\frac{7}{11} \times \frac{8}{10} =$ _____

13. $\frac{2}{11} \times \frac{4}{5} =$ _____

14. $\frac{7}{10} \times \frac{2}{4} =$ _____

15. $\frac{5}{10} \times \frac{4}{7} =$ _____

16. $\frac{9}{10} \times \frac{4}{12} =$ _____

17. $\frac{4}{9} \times \frac{11}{12} =$ _____

18. $\frac{2}{5} \times \frac{1}{11} =$ _____

19. $\frac{1}{5} \times \frac{2}{12} =$ _____

20. $\frac{6}{12} \times \frac{6}{8} =$ _____

21. $\frac{1}{8} \times \frac{7}{12} =$ _____

Directions: Solve each word problem below.

22. Jason had 24 gumballs. He gave 2/3 of the gumballs to Marcy. He gave 3/4 of the remaining gumballs to David. How many gumballs did Jason give away? How many does he have left?

23. Penelope had 36 yo-yos. She sold 3/9 of them to the local yo-yo shop. She sold 1/2 of the remaining yo-yos at a garage sale. How many yo-yos did Penelope sell? How many yo-yos does she have left?

Practice 24 ⟳ ☙ ⟳ ☙ ⟳ ☙ ⟳ ☙ ⟳ ☙ ⟳ ☙ ⟳ ⟳ ☙

> To divide fractions:
>
> 1. Flip the numerator and the denominator in the second fraction.
> 2. Multiply the numerators.
> 3. Multiply the denominators.
> 4. Reduce the answer to its simplest form.
>
> For example: $\dfrac{5}{12} \div \dfrac{3}{11} = \dfrac{5}{12} \times \dfrac{11}{3} = \dfrac{55}{36} = \dfrac{119}{36}$

Directions: Solve the problems below.

1. $\dfrac{6}{11} \div \dfrac{3}{6}$ 2. $\dfrac{7}{11} \div \dfrac{1}{9}$ 3. $\dfrac{5}{9} \div \dfrac{4}{10}$ 4. $\dfrac{3}{5} \div \dfrac{7}{10}$

5. $\dfrac{6}{10} \div \dfrac{5}{6}$ 6. $\dfrac{4}{12} \div \dfrac{7}{11}$ 7. $\dfrac{9}{10} \div \dfrac{5}{10}$ 8. $\dfrac{1}{11} \div \dfrac{5}{6}$

9. $\dfrac{1}{6} \div \dfrac{2}{5}$ 10. $\dfrac{7}{8} \div \dfrac{9}{11}$ 11. $\dfrac{2}{10} \div \dfrac{2}{7}$ 12. $\dfrac{10}{11} \div \dfrac{7}{9}$

13. $\dfrac{1}{4} \div \dfrac{1}{8}$ 14. $\dfrac{2}{4} \div \dfrac{9}{12}$ 15. $\dfrac{4}{8} \div \dfrac{6}{8}$ 16. $\dfrac{11}{12} \div \dfrac{3}{4}$

17. $\dfrac{3}{7} \div \dfrac{5}{7}$ 18. $\dfrac{4}{5} \div \dfrac{6}{9}$ 19. $\dfrac{1}{2} \div \dfrac{5}{11}$ 20. $\dfrac{3}{6} \div \dfrac{8}{10}$

Practice 25 ৯ ৩ ৯ ৩ ৯ ৩ ৯ ৩ ৯ ৩ ৯ ৩ ৯ ৩ ৯ ৩ ৯

> To divide two fractions:
>
> 1. Flip the numerator and the denominator in the second fraction.
> 2. Multiply the numerators.
> 3. Multiply the denominators.
> 4. Reduce the answer to its simplest form.
>
> For example: $\frac{5}{12} \div \frac{7}{8} = \frac{5}{12} \times \frac{8}{7} = \frac{40}{84} = \frac{10}{21}$

Directions: Solve the problems below.

1. $\frac{2}{12} \div \frac{2}{10} =$ _____

2. $\frac{5}{6} \div \frac{1}{8} =$ _____

3. $\frac{6}{7} \div \frac{2}{9} =$ _____

4. $\frac{3}{4} \div \frac{8}{9} =$ _____

5. $\frac{1}{7} \div \frac{3}{6} =$ _____

6. $\frac{1}{9} \div \frac{5}{11} =$ _____

7. $\frac{3}{10} \div \frac{9}{12} =$ _____

8. $\frac{5}{8} \div \frac{6}{9} =$ _____

9. $\frac{4}{10} \div \frac{10}{11} =$ _____

10. $\frac{6}{11} \div \frac{8}{11} =$ _____

11. $\frac{8}{12} \div \frac{6}{7} =$ _____

12. $\frac{8}{10} \div \frac{6}{11} =$ _____

13. $\frac{4}{8} \div \frac{2}{9} =$ _____

14. $\frac{2}{12} \div \frac{4}{6} =$ _____

15. $\frac{1}{7} \div \frac{2}{11} =$ _____

16. $\frac{11}{12} \div \frac{8}{11} =$ _____

17. $\frac{6}{8} \div \frac{9}{10} =$ _____

18. $\frac{3}{4} \div \frac{1}{4} =$ _____

19. $\frac{5}{6} \div \frac{2}{5} =$ _____

20. $\frac{2}{9} \div \frac{4}{6} =$ _____

Directions: Solve the following problems. Circle the problem in each pair that has the larger answer.

21. $\frac{1}{4} \times \frac{1}{4} =$ _____

22. $\frac{1}{2} \times \frac{1}{2} =$ _____

 $\frac{1}{4} \div \frac{1}{4} =$ _____

 $\frac{1}{2} \div \frac{1}{2} =$ _____

23. $\frac{3}{4} \times \frac{1}{2} =$ _____

24. $\frac{5}{6} \times \frac{1}{9} =$ _____

 $\frac{3}{4} \div \frac{1}{2} =$ _____

 $\frac{5}{6} \div \frac{1}{9} =$ _____

Fraction Sets

Practice 26

Directions: Use multiplication to find the fractional amount for each set below.

For example: $\frac{1}{4}$ of $8 = \frac{1}{4} \times \frac{8}{1} = \frac{8}{4} = 2$

1. $\frac{2}{3}$ of 12	**2.** $\frac{6}{7}$ of 14	**3.** $\frac{4}{5}$ of 5 =
4. $\frac{4}{5}$ of 20 =	**5.** $\frac{3}{4}$ of 20 =	**6.** $\frac{1}{3}$ of 18 =
7. $\frac{7}{9}$ of 45	**8.** $\frac{9}{12}$ of 96	**9.** $\frac{8}{10}$ of 50
10. $\frac{1}{10}$ of 60	**11.** $\frac{8}{9}$ of 63	**12.** $\frac{3}{4}$ of 16
13. $\frac{1}{9}$ of 81	**14.** $\frac{3}{6}$ of 60	**15.** $\frac{4}{9}$ of 45
16. $\frac{4}{6}$ of 36	**17.** $\frac{2}{10}$ of 40	**18.** $\frac{1}{12}$ of 36

Practice 27 ➋ ➌ ➋ ➌ ➋ ➌ ➋ ➌ ➋ ➌ ➋ ➌ ➋ ➌ ➋

Directions: Use multiplication to find the fractional amount for each set below.

For example: $\frac{1}{4}$ of $8 = \frac{1}{4} \times \frac{8}{1} = \frac{8}{4} = 2$

1. $\frac{1}{10}$ of 90	**2.** $\frac{2}{5}$ of 25	**3.** $\frac{5}{8}$ of 40
4. $\frac{2}{7}$ of 84	**5.** $\frac{2}{9}$ of 45	**6.** $\frac{1}{6}$ of 48
7. $\frac{9}{10}$ of 30	**8.** $\frac{6}{7}$ of 63	**9.** $\frac{3}{5}$ of 75
10. $\frac{1}{2}$ of 46	**11.** $\frac{5}{9}$ of 27	**12.** $\frac{5}{6}$ of 30
13. $\frac{1}{8}$ of 48	**14.** $\frac{1}{4}$ of 68	**15.** $\frac{7}{9}$ of 72
16. $\frac{4}{7}$ of 21	**17.** $\frac{3}{10}$ of 70	**18.** $\frac{3}{4}$ of 20

Practice 28

Directions: Solve each problem. Circle the corresponding letter of the correct answer.

1. Which is more?

A. $\frac{4}{7}$ of 28

B. $\frac{3}{4}$ of 28

C. $1\frac{1}{4}$ of 28

2. Which is more?

A. $\frac{2}{6}$ of 18

B. $\frac{1}{2}$ of 18

C. $\frac{5}{9}$ of 18

3. Which is more?

A. $\frac{2}{3}$ of 24

B. $\frac{5}{6}$ of 24

C. $\frac{3}{8}$ of 24

4. Which is more?

A. $\frac{2}{9}$ of 36

B. $\frac{5}{6}$ of 36

C. $\frac{1}{4}$ of 36

5. Which is more?

A. $\frac{8}{12}$ of 36

B. $\frac{1}{3}$ of 36

C. $\frac{3}{6}$ of 36

6. Which is more?

A. $\frac{7}{8}$ of 16

B. $\frac{3}{4}$ of 16

C. $\frac{1}{2}$ of 16

7. Which is more?

A. $\frac{3}{4}$ of 40

B. $\frac{4}{10}$ of 40

C. $\frac{3}{8}$ of 40

8. Which is more?

A. $\frac{6}{7}$ of 35

B. $\frac{3}{5}$ of 35

C. $\frac{12}{35}$ of 35

9. Which is more?

A. $\frac{7}{12}$ of 60

B. $\frac{8}{10}$ of 60

C. $\frac{2}{6}$ of 60

10. Which is more?

A. $\frac{1}{4}$ of 20

B. $\frac{2}{5}$ of 20

C. $\frac{3}{10}$ of 20

11. Which is more?

A. $\frac{3}{7}$ of 63

B. $\frac{1}{9}$ of 63

C. $\frac{2}{3}$ of 63

12. Which is more?

A. $\frac{2}{4}$ of 30

B. $\frac{1}{3}$ of 30

C. $\frac{1}{10}$ of 30

Practice 29 ∂ ℮ ∂ ℮ ∂ ℮ ∂ ℮ ∂ ℮ ∂ ∂ ℮

Directions: Solve each problem. Then circle the corresponding letter of the correct answer.

1. Which is less?

   A. $\frac{1}{3}$ of 12

   B. $\frac{1}{4}$ of 12

   C. $\frac{1}{2}$ of 12

2. Which is less?

   A. $\frac{1}{6}$ of 36

   B. $\frac{2}{9}$ of 36

   C. $\frac{1}{3}$ of 36

3. Which is less?

   A. $\frac{3}{4}$ of 20

   B. $\frac{2}{5}$ of 20

   C. $\frac{7}{10}$ of 20

4. Which is less?

   A. $\frac{1}{8}$ of 24

   B. $\frac{5}{6}$ of 24

   C. $\frac{3}{4}$ of 24

5. Which is less?

   A. $\frac{1}{2}$ of 16

   B. $\frac{3}{8}$ of 16

   C. $\frac{3}{4}$ of 16

6. Which is less?

   A. $\frac{3}{5}$ of 50

   B. $\frac{8}{10}$ of 50

   C. $\frac{1}{2}$ of 50

7. Which is less?

   A. $\frac{9}{10}$ of 80

   B. $\frac{3}{8}$ of 80

   C. $\frac{1}{4}$ of 80

8. Which is less?

   A. $\frac{5}{9}$ of 36

   B. $\frac{4}{6}$ of 36

   C. $\frac{5}{12}$ of 36

9. Which is less?

   A. $\frac{3}{10}$ of 10

   B. $\frac{1}{2}$ of 10

   C. $\frac{2}{5}$ of 10

10. Which is less?

   A. $\frac{6}{7}$ of 28

   B. $\frac{3}{4}$ of 28

   C. $\frac{1}{2}$ of 28

11. Which is less?

   A. $\frac{1}{4}$ of 32

   B. $\frac{3}{8}$ of 32

   C. $\frac{5}{16}$ of 32

12. Which is less?

   A. $\frac{2}{3}$ of 18

   B. $\frac{1}{9}$ of 18

   C. $\frac{4}{6}$ of 18

Practice 30 🐚 🐚 🐚 🐚 🐚 🐚 🐚 🐚 🐚 🐚 🐚 🐚 🐚 🐚

Directions: Find the decimal equivalent for each fraction by dividing the numerator by the denominator. Write the answer to two places to the right of the decimal point. Then use a calculator to check your work.

For example: $\frac{10}{11} = 10 \div 11 = .09$

1. $\frac{2}{3}$	2. $\frac{2}{4}$	3. $\frac{3}{4}$	4. $\frac{9}{10}$
5. $\frac{2}{5}$	6. $\frac{1}{8}$	7. $\frac{1}{10}$	8. $\frac{1}{12}$
9. $\frac{4}{5}$	10. $\frac{1}{3}$	11. $\frac{1}{9}$	12. $\frac{1}{6}$
13. $\frac{5}{6}$	14. $\frac{3}{10}$	15. $\frac{1}{5}$	16. $\frac{5}{10}$
17. $\frac{3}{7}$	18. $\frac{1}{4}$	19. $\frac{3}{5}$	20. $\frac{6}{11}$

Practice 31 ᓂ ᘓ ᓂ ᘓ ᓂ ᘓ ᓂ ᘓ ᓂ ᘓ ᓂ ᘓ ᓂ ᓂ ᘓ

- To change a fraction to a decimal, divide the numerator by the denominator.
- To change a decimal to a fraction, multiply the decimal by 100%.

Directions: Complete the chart below. Write each decimal answer to the hundredths place, two places to the right of the decimal point.

Fraction	Decimal	Percent
1. $\frac{1}{2}$	$1 \div 2 = .50$	$.50 \times 100\% = 50\%$
2. $\frac{1}{3}$		
3. $\frac{2}{3}$		
4. $\frac{1}{4}$		
5. $\frac{3}{4}$		
6. $\frac{1}{5}$		
7. $\frac{2}{5}$		
8. $\frac{3}{5}$		
9. $\frac{4}{5}$		
10. $\frac{1}{6}$		
11. $\frac{5}{6}$		
12. $\frac{1}{8}$		
13. $\frac{5}{8}$		
14. $\frac{7}{8}$		
15. $\frac{1}{9}$		

Practice 32 ❧ ❧ ❧ ❧ ❧ ❧ ❧ ❧ ❧ ❧ ❧ ❧ ❧ ❧

Directions: Solve each problem and write the answer as a fraction, a decimal, and a percent.

1. Roger answered 4 out of 5 questions correctly.

   Fraction: _____

   Decimal: _____

   Percent: _____

2. Gretchen finished 3 out of the 4 chores.

   Fraction: _____

   Decimal: _____

   Percent: _____

3. Selina hit 1 out of 6 pitches during practice.

   Fraction: _____

   Decimal: _____

   Percent: _____

4. Bobby found 1 of his 2 missing pairs of socks.

   Fraction: _____

   Decimal: _____

   Percent: _____

5. Jane ate 3 of the 7 pieces of candy.

   Fraction: _____

   Decimal: _____

   Percent: _____

6. Jason broke 1 of the 3 glasses.

   Fraction: _____

   Decimal: _____

   Percent: _____

7. Tamra skied 7 of the 10 slopes.

   Fraction: _____

   Decimal: _____

   Percent: _____

8. Brian built 3 of the 5 model airplanes.

   Fraction: _____

   Decimal: _____

   Percent: _____

9. Laura ate 4 of the 7 cookies.

   Fraction: _____

   Decimal: _____

   Percent: _____

10. Brett won 2 of the 3 sporting events.

    Fraction: _____

    Decimal: _____

    Percent: _____

11. Marsha popped 1 of the 8 balloons.

    Fraction: _____

    Decimal: _____

    Percent: _____

12. Out of every 9 flights, 2 are on time.

    Fraction: _____

    Decimal: _____

    Percent: _____

13. Sabrina saves $1 out of every $4 she earns.

    Fraction: _____

    Decimal: _____

    Percent: _____

14. Of 10 movies Ben directed, only 1 was a hit.

    Fraction: _____

    Decimal: _____

    Percent: _____

15. Of the 9 cars in the parking lot, 4 are green.

    Fraction: _____

    Decimal: _____

    Percent: _____

Practice 33 ꙮ ꙮ ꙮ ꙮ ꙮ ꙮ ꙮ ꙮ ꙮ ꙮ ꙮ ꙮ ꙮ ꙮ

A benchmark is a commonly used amount or quantity used for comparison, such as one-half or one-third. A fraction or decimal can be compared against a given benchmark to determine if it is more than, the same, or less than the benchmark.

For a fraction with a benchmark of $\frac{1}{2}$ or .5:

- If the numerator is exactly one-half of the denominator, then the fraction is equal to

  $\frac{1}{2}$ or .50. For example: $\frac{3}{6}, \frac{4}{8}, \frac{6}{12}$

- If the numerator is less than one-half of the denominator, then the fraction is less than

  $\frac{1}{2}$ or .5. For example: $\frac{1}{4}, \frac{2}{5}, \frac{4}{9}$

- If the numerator is greater than one-half of the denominator, then the fraction is greater than

  $\frac{1}{2}$ or .5. For example: $\frac{2}{3}, \frac{5}{7}, \frac{9}{10}$

Directions: Compare the following amounts to the benchmark of .5. Write the number that is one-half of the denominator. Then circle whether the fraction is < (less than), > (greater than), or = (equal to) the benchmark.

Fraction	Half of the Denominator	Benchmark	Fraction	Half of the Denominator	Benchmark
1. $\frac{2}{5}$		< .5 = .5 > .5	9. $\frac{3}{5}$		< .5 = .5 > .5
2. $\frac{7}{12}$		< .5 = .5 > .5	10. $\frac{1}{4}$		< .5 = .5 > .5
3. $\frac{5}{6}$		< .5 = .5 > .5	11. $\frac{6}{7}$		< .5 = .5 > .5
4. $\frac{1}{3}$		< .5 = .5 > .5	12. $\frac{1}{2}$		< .5 = .5 > .5
5. $\frac{1}{6}$		< .5 = .5 > .5	13. $\frac{3}{4}$		< .5 = .5 > .5
6. $\frac{2}{3}$		< .5 = .5 > .5	14. $\frac{1}{11}$		< .5 = .5 > .5
7. $\frac{8}{11}$		< .5 = .5 > .5	15. $\frac{5}{8}$		< .5 = .5 > .5
8. $\frac{7}{9}$		< .5 = .5 > .5	16. $\frac{5}{7}$		< .5 = .5 > .5

Practice 34 ⟳ ❧ ⟳ ❧ ⟳ ❧ ⟳ ❧ ⟳ ❧ ⟳ ❧ ⟳ ⟳ ❧

A benchmark is a commonly used amount or quantity used for comparison, such as one-half or one-third. A fraction or decimal can be compared against a given benchmark to determine if it is more than, the same, or less than the benchmark.

For a fraction with a benchmark of $\frac{1}{3}$ or .33:

- If the numerator is one-third of the denominator, then the fraction is equal to $\frac{1}{3}$ or .33.
  For example: $\frac{2}{6}, \frac{3}{9}, \frac{4}{12}$
- If the numerator is less than one-third of the denominator, then the fraction is less than .33.
  For example: $\frac{1}{4}, \frac{2}{7}, \frac{3}{11}$
- If the numerator is greater than one-third of the denominator, then the fraction is greater than .33. For example: $\frac{2}{3}, \frac{5}{7}, \frac{9}{10}$

Directions: Compare the following amounts to the benchmark of .33. Write the number that is about one-third of the denominator. Then circle whether the fraction is < (less than), > (greater than), or = (equal to) the benchmark.

For a fraction with a benchmark of $\frac{1}{4}$ or .25:

- If the numerator is one-fourth of the denominator, then the fraction is equal to $\frac{1}{4}$ or .25.
  For example: $\frac{2}{8}, \frac{3}{12}, \frac{4}{16}$
- If the numerator is less than one-fourth of the denominator, then the fraction is less than $\frac{1}{4}$ or .25. For example: $\frac{1}{8}, \frac{2}{9}, \frac{3}{14}$
- If the numerator is greater than one-fourth of the denominator, then the fraction is greater than $\frac{1}{4}$ or .25. For example: $\frac{2}{3}, \frac{5}{7}, \frac{9}{10}$

Directions: Compare the following amounts to the benchmark of .25. Write the number that is about one-fourth of the denominator. Then circle whether the fraction is < (less than), > (greater than), or = (equal to) the benchmark.

Fraction	$\frac{1}{3}$ of the Denominator	Benchmark
1. $\frac{7}{11}$		< .33 = .33 > .33
2. $\frac{1}{5}$		< .33 = .33 > .33
3. $\frac{7}{8}$		< .33 = .33 > .33
4. $\frac{3}{9}$		< .33 = .33 > .33
5. $\frac{8}{9}$		< .33 = .33 > .33
6. $\frac{3}{11}$		< .33 = .33 > .33
7. $\frac{1}{8}$		< .33 = .33 > .33

Fraction	$\frac{1}{4}$ of the Denominator	Benchmark
8. $\frac{2}{8}$		< .25 = .25 > .25
9. $\frac{1}{7}$		< .25 = .25 > .25
10. $\frac{3}{23}$		< .25 = .25 > .25
11. $\frac{4}{9}$		< .25 = .25 > .25
12. $\frac{2}{9}$		< .25 = .25 > .25
13. $\frac{3}{10}$		< .25 = .25 > .25
14. $\frac{9}{11}$		< .25 = .25 > .25

Practice 35

Directions: Use the symbols > (greater than), < (less than), or = (equal to) to compare each pair of numbers below.

1.	$\frac{2}{5}$ ◯ .30		**2.**	$\frac{4}{5}$ ◯ .75		**3.**	$\frac{9}{10}$ ◯ .08
4.	$\frac{3}{7}$ ◯ .10		**5.**	$\frac{1}{9}$ ◯ .20		**6.**	$\frac{1}{12}$ ◯ .50
7.	$\frac{1}{3}$ ◯ .83		**8.**	$\frac{2}{3}$ ◯ .42		**9.**	$\frac{3}{4}$ ◯ .90
10.	$\frac{5}{10}$ ◯ .50		**11.**	$\frac{1}{5}$ ◯ .50		**12.**	$\frac{1}{4}$ ◯ .80
13.	$\frac{1}{5}$ ◯ .12		**14.**	$\frac{5}{6}$ ◯ .11		**15.**	$\frac{1}{10}$ ◯ .66
16.	$\frac{2}{4}$ ◯ .16		**17.**	$\frac{1}{8}$ ◯ .40		**18.**	$\frac{3}{10}$ ◯ .33

Directions: Write the equivalent fractions for the following decimal amounts.

19. .50 _____   **20.** .25 _____   **21.** .33 _____

22. .75 _____   **23.** .20 _____   **24.** .66 _____

25. .40 _____   **26.** .60 _____   **27.** .80 _____

Practice 36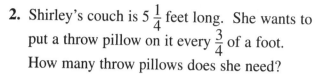

Directions: Solve each word problem below

1. A large, submarine sandwich was $6\frac{2}{3}$ feet long. It was cut into pieces $\frac{5}{6}$ of a foot in length. How many smaller sandwiches were there?

_____ sandwiches

2. Shirley's couch is $5\frac{1}{4}$ feet long. She wants to put a throw pillow on it every $\frac{3}{4}$ of a foot. How many throw pillows does she need?

_____ throw pillows

3. George bought a $1\frac{1}{3}$ foot-long hot dog. He cut it into pieces $\frac{2}{3}$ of a foot in length. How many pieces did he have?

_____ pieces of hot dog

4. Jed needs to cut a $2\frac{1}{2}$ foot board into pieces $\frac{1}{4}$ of a foot in length. How many boards will he have then?

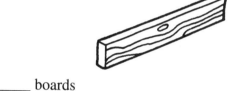

_____ boards

5. Kathy bought $3\frac{2}{3}$ yards of fabric to make place mats. It takes $\frac{1}{6}$ of a yard of fabric for each place mat. How many place mats can she make?

_____ place mats

6. Heidi had a piece of ribbon that was $8\frac{1}{3}$ feet long. She cut it into sections $\frac{1}{3}$ of a foot in length. How many pieces did Heidi have?

_____ pieces of ribbon

7. A $4\frac{1}{5}$ foot long piece of poster board was cut into pieces $\frac{3}{10}$ of a foot in length. How many pieces of poster board are there?

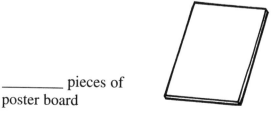

_____ pieces of poster board

8. A $1\frac{1}{2}$ foot long rope was cut into pieces $\frac{1}{8}$ of a foot in length. How many pieces were there?

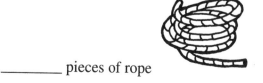

_____ pieces of rope

Test Practice 1

Directions: Solve each problem below. Use the Answer Sheet to fill in the correct answers.

1. What fraction of the circle is shaded?

$\frac{1}{8}$ $\frac{7}{8}$ $\frac{6}{7}$
Ⓐ   Ⓑ   Ⓒ

2. Name the fraction.

$\frac{5}{9}$ $\frac{4}{5}$ $\frac{4}{9}$
Ⓐ   Ⓑ   Ⓒ

3. What fraction of the rectangle is shaded?

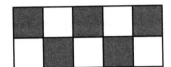

$\frac{5}{10}$ $\frac{5}{5}$ $\frac{5}{1}$
Ⓐ   Ⓑ   Ⓒ

4. Shade each circle as marked. Then compare the fractions.

$\frac{2}{3}$ _____ $\frac{1}{7}$

>   <   =
Ⓐ   Ⓑ   Ⓒ

5. Shade each circle as marked. Then compare the fractions.

$\frac{3}{5}$ _____ $\frac{5}{6}$

>   <   =
Ⓐ   Ⓑ   Ⓒ

6. Shade each circle as marked. Then compare the fractions.

$\frac{5}{9}$ _____ $\frac{5}{8}$

>   <   =
Ⓐ   Ⓑ   Ⓒ

7. Identify the fraction shown by the bold line.

0   1

$\frac{1}{2}$ $\frac{1}{3}$ $\frac{1}{4}$
Ⓐ   Ⓑ   Ⓒ

8. Identify the fraction shown by the bold line.

0   1

$\frac{1}{4}$ $\frac{2}{3}$ $\frac{3}{4}$
Ⓐ   Ⓑ   Ⓒ

9. Add.

$$\frac{3}{8} + \frac{2}{8}$$

$\frac{5}{5}$ $\frac{5}{8}$ $\frac{1}{8}$
Ⓐ   Ⓑ   Ⓒ

10. Subtract.

$$\frac{8}{9} - \frac{3}{9}$$

$\frac{5}{9}$ $\frac{4}{9}$ $\frac{5}{4}$
Ⓐ   Ⓑ   Ⓒ

11. Find the common denominator.

$$\frac{6}{8} + \frac{2}{7}$$

15   56   24
Ⓐ   Ⓑ   Ⓒ

12. Find the common denominator.

$$\frac{3}{5} + \frac{1}{4}$$

9   15   20
Ⓐ   Ⓑ   Ⓒ

Test Practice 2

Directions: Solve each problem below. Use the Answer Sheet to fill in the correct answers.

---

1. Find the common denominator.

$$\frac{3}{5} \text{ and } \frac{4}{7}$$

12	15	35
Ⓐ	Ⓑ	Ⓒ

2. Find the common denominator.

$$\frac{1}{6} \text{ and } \frac{1}{9}$$

36	15	45
Ⓐ	Ⓑ	Ⓒ

3. Reduce to its simplest form.

$$\frac{2}{8}$$

$\frac{2}{4}$	$\frac{1}{4}$	$\frac{3}{4}$
Ⓐ	Ⓑ	Ⓒ

---

4. Reduce to its simplest form.

$$\frac{8}{12}$$

46	$\frac{1}{4}$	$\frac{2}{3}$
Ⓐ	Ⓑ	Ⓒ

5. Add.

$$\frac{1}{6} + \frac{4}{9}$$

$\frac{5}{15}$	$\frac{11}{18}$	$\frac{10}{15}$
Ⓐ	Ⓑ	Ⓒ

6. Add.

$$\frac{1}{4} + \frac{1}{2}$$

$\frac{2}{4}$	$\frac{3}{4}$	$\frac{1}{8}$
Ⓐ	Ⓑ	Ⓒ

---

7. Subtract.

$$\frac{7}{8} - \frac{2}{3}$$

$\frac{5}{24}$	$\frac{5}{5}$	$\frac{14}{12}$
Ⓐ	Ⓑ	Ⓒ

8. Subtract.

$$\frac{3}{5} - \frac{1}{3}$$

$\frac{4}{15}$	$\frac{2}{5}$	$\frac{2}{3}$
Ⓐ	Ⓑ	Ⓒ

9. Find the equivalent fraction.

$$\frac{1}{4}$$

$\frac{1}{8}$	$\frac{2}{4}$	$\frac{2}{8}$
Ⓐ	Ⓑ	Ⓒ

---

10. Find the equivalent fraction.

$$\frac{1}{5}$$

$\frac{2}{5}$	$\frac{2}{10}$	$\frac{1}{10}$
Ⓐ	Ⓑ	Ⓒ

11. Identify the fraction shown by the bold line.

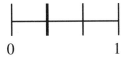

$\frac{2}{6}$	$\frac{1}{3}$	$\frac{2}{3}$
Ⓐ	Ⓑ	Ⓒ

12. Name the missing fraction.

$\frac{1}{6}$	$\frac{2}{6}$	$\frac{1}{3}$
Ⓐ	Ⓑ	Ⓒ

Test Practice 3 ⟳ ⟲ ⟳ ⟲ ⟳ ⟲ ⟳ ⟲ ⟳ ⟳ ⟲

Directions: Solve each problem below. Use the Answer Sheet to fill in the correct answers.

1. Identify the mixed fraction shown by the bold line.

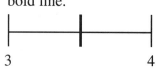

3   4

$3\frac{1}{4}$ (A) $3\frac{1}{2}$ (B) $3\frac{2}{3}$ (C)

2. Identify the mixed fraction shown by the bold line.

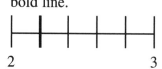

2   3

$2\frac{1}{4}$ (A) $2\frac{1}{5}$ (B) $2\frac{1}{6}$ (C)

3. Add.

$$3\frac{1}{5} + 6\frac{3}{5}$$

$9\frac{4}{5}$ (A) $3\frac{2}{5}$ (B) $12\frac{3}{5}$ (C)

---

4. Subtract.

$$9\frac{7}{8} - 7\frac{3}{8}$$

$2\frac{1}{2}$ (A) $2\frac{10}{8}$ (B) $2\frac{4}{16}$ (C)

5. Write the amount in standard form.

$$\$8\frac{5}{100}$$

$8.5000   (A) $8.50   (B) $8.05   (C)

6. Write the amount in standard form.

$$\$3\frac{29}{100}$$

$3.29   (A) $3.029   (B) $329.00   (C)

---

7. Rewrite the amount as a fraction.

$.88

$\$\frac{8}{100}$ (A) $\$\frac{88}{100}$ (B) $\$\frac{100}{88}$ (C)

8. Rewrite the amount as a mixed fraction.

$2.74

$\$\frac{274}{00}$ (A) $\$27\frac{4}{100}$ (B) $\$2\frac{74}{100}$ (C)

9. Subtract.

$$\begin{array}{r} \$9.36 \\ -\ \$4.29 \\ \hline \end{array}$$

$4.07   (A) $5.13   (B) $5.07   (C)

---

10. Add.

$$\begin{array}{r} \$1.86 \\ +\ \$12.04 \\ \hline \end{array}$$

$13.90   (A) $12.90   (B) $13.00   (C)

11. Change the improper fraction to a mixed fraction.

$$\frac{48}{5}$$

$9\frac{3}{5}$ (A) $9\frac{8}{5}$ (B) $9\frac{4}{5}$ (C)

12. Change the improper fraction to a mixed fraction.

$$\frac{33}{10}$$

$3\frac{33}{10}$ (A) $3\frac{3}{10}$ (B) $3\frac{1}{10}$ (C)

Test Practice 4

Directions: Solve each problem below. Use the Answer Sheet to fill in the correct answers.

1. Change the mixed fraction to an improper fraction.

$$2\frac{3}{8}$$

$\frac{16}{8}$ Ⓐ $\frac{6}{8}$ Ⓑ $\frac{19}{8}$ Ⓒ

2. Change the mixed fraction to an improper fraction.

$$4\frac{1}{6}$$

$\frac{4}{6}$ Ⓐ $\frac{25}{6}$ Ⓑ $\frac{24}{6}$ Ⓒ

3. Add. Rewrite as a mixed fraction.

$$\frac{6}{5}+\frac{7}{5}$$

$2\frac{3}{5}$ Ⓐ $\frac{13}{5}$ Ⓑ $\frac{13}{10}$ Ⓒ

4. Subtract. Rewrite as a mixed fraction.

$$\frac{13}{3}-\frac{9}{3}$$

$7\frac{1}{3}$ Ⓐ $\frac{4}{3}$ Ⓑ $1\frac{1}{3}$ Ⓒ

5. Compare.

$$9\frac{1}{4} \rule{2cm}{0.4pt} \frac{69}{8}$$

$>$ Ⓐ $<$ Ⓑ $=$ Ⓒ

6. Compare.

$$3\frac{1}{2} \rule{2cm}{0.4pt} \frac{12}{5}$$

$>$ Ⓐ $<$ Ⓑ $=$ Ⓒ

7. Subtract.

$$4\frac{4}{5}-\frac{18}{5}$$

$\frac{6}{5}$ Ⓐ $\frac{24}{5}$ Ⓑ $\frac{14}{5}$ Ⓒ

8. Add.

$$7\frac{1}{17}+\frac{35}{7}$$

$\frac{42}{7}$ Ⓐ $12\frac{1}{17}$ Ⓑ $\frac{28}{7}$ Ⓒ

9. Locate $7\frac{4}{6}$ on the number line.

```
  |--+--+--+--+--+--|
  7   A  B  C      8
```

A Ⓐ B Ⓑ C Ⓒ

10. Plot $5\frac{1}{4}$ on the number line.

```
  |--+--+--+--|
  5  A  B  C  6
```

A Ⓐ B Ⓑ C Ⓒ

11. Plot $3\frac{4}{5}$ on the number line.

```
  |--+--+--+--+--|
  3  A     B  C  4
```

A Ⓐ B Ⓑ C Ⓒ

12. Multiply. Reduce the answer.

$$\frac{2}{3} \times \frac{3}{4}$$

$\frac{1}{2}$ Ⓐ $\frac{6}{7}$ Ⓑ $\frac{8}{9}$ Ⓒ

Test Practice 5 ⌒⌒⌒⌒⌒⌒⌒⌒⌒⌒⌒⌒

Directions: Solve each problem below. Use the Answer Sheet to fill in the correct answer.

---

1. Multiply.

$$\frac{5}{6} \times \frac{1}{7}$$

$\frac{30}{5}$ $\frac{5}{42}$ $\frac{42}{5}$

(A)   (B)   (C)

2. Multiply.

$$\frac{4}{5} \times \frac{1}{6}$$

$\frac{2}{15}$ $\frac{4}{11}$ $\frac{20}{6}$

(A)   (B)   (C)

3. Multiply.

$$\frac{5}{6} \times \frac{7}{12}$$

$\frac{35}{12}$ $\frac{12}{18}$ $\frac{35}{72}$

(A)   (B)   (C)

---

4. Divide. Reduce the answer.

$$\frac{2}{9} \div \frac{4}{9}$$

$\frac{8}{9}$ $\frac{1}{2}$ $\frac{2}{9}$

(A)   (B)   (C)

5. Divide. Reduce the answer.

$$\frac{4}{12} \div \frac{3}{10}$$

$1\frac{1}{9}$ $\frac{12}{120}$ $12\frac{1}{10}$

(A)   (B)   (C)

6. Divide. Reduce the answer.

$$\frac{4}{12} \div \frac{4}{10}$$

$\frac{6}{5}$ $\frac{5}{6}$ $\frac{2}{15}$

(A)   (B)   (C)

---

7. Find the answer.

$$\frac{1}{7} \times 35$$

5   7   8

(A)   (B)   (C)

8. Find the answer.

$$\frac{2}{3} \times 21$$

4   14   6

(A)   (B)   (C)

9. Find the answer.

$$\frac{7}{8} \times 24$$

21   14   32

(A)   (B)   (C)

---

10. Which is more?

(A) $\frac{3}{4}$ of 12

(B) $\frac{1}{2}$ of 12

(C) $\frac{5}{6}$ of 12

11. Which is more?

(A) $\frac{5}{7}$ of 12

(B) $\frac{2}{3}$ of 12

(C) $\frac{5}{6}$ of 12

12. Which is less?

(A) $\frac{1}{2}$ of 18

(B) $\frac{5}{6}$ of 18

(C) $\frac{1}{3}$ of 18

Test Practice 6 🐚 🌀 🐚 🌀 🐚 🌀 🐚 🌀 🐚 🐚 🌀

Directions: Solve each problem below. Use the Answer Sheet to fill in the correct answer.

1. Identify the percent.	**2.** Identify the decimal.	**3.** Which number is .50?
$\dfrac{4}{5}$	$\dfrac{4}{5} \times \dfrac{1}{6}$	
50% 40% 80%	.34 .13 3.4	$\dfrac{3}{4}$ $\dfrac{2}{8}$ $\dfrac{3}{6}$
Ⓐ Ⓑ Ⓒ	Ⓐ Ⓑ Ⓒ	Ⓐ Ⓑ Ⓒ

4. Which number is > .33?	**5.** Compare the two amounts.	**6.** Compare the two amounts.
	$\dfrac{2}{5}$ _____ .10	$\dfrac{1}{4}$ _____ .66
$\dfrac{2}{6}$ $\dfrac{1}{4}$ $\dfrac{2}{3}$	> < =	> < =
Ⓐ Ⓑ Ⓒ	Ⓐ Ⓑ Ⓒ	Ⓐ Ⓑ Ⓒ

7. A rope is $2\frac{1}{4}$ feet long. If it is cut into pieces $\frac{3}{4}$ of a foot in length, how many pieces will there be?	**8.** A machine can make $\frac{2}{3}$ yard of chain-link fence in an hour. How many yards can be made in $\frac{1}{4}$ of an hour?	**9.** Add.
		$\dfrac{7}{10} + \dfrac{1}{9}$
2 3 4	$\dfrac{8}{3}$ $\dfrac{1}{6}$ $\dfrac{3}{7}$	$\dfrac{7}{19}$ $\dfrac{73}{90}$ $\dfrac{6}{9}$
Ⓐ Ⓑ Ⓒ	Ⓐ Ⓑ Ⓒ	Ⓐ Ⓑ Ⓒ

10. Solve the problem.	**11.** Multiply. Simplify the answer.	**12.** Divide. Write the answer as a mixed fraction.
$\dfrac{9}{10} - \dfrac{1}{5}$	$\dfrac{1}{6} \times \dfrac{3}{8}$	$\dfrac{4}{5} \div \dfrac{3}{6}$
$\dfrac{7}{10}$ $\dfrac{10}{5}$ $\dfrac{8}{5}$	$\dfrac{1}{16}$ $\dfrac{3}{14}$ $\dfrac{3}{24}$	$\dfrac{8}{5}$ $1\dfrac{3}{5}$ $1\dfrac{1}{6}$
Ⓐ Ⓑ Ⓒ	Ⓐ Ⓑ Ⓒ	Ⓐ Ⓑ Ⓒ

Answer Sheet

Test Practice 1

1. Ⓐ Ⓑ Ⓒ Ⓓ
2. Ⓐ Ⓑ Ⓒ Ⓓ
3. Ⓐ Ⓑ Ⓒ Ⓓ
4. Ⓐ Ⓑ Ⓒ Ⓓ
5. Ⓐ Ⓑ Ⓒ Ⓓ
6. Ⓐ Ⓑ Ⓒ Ⓓ
7. Ⓐ Ⓑ Ⓒ Ⓓ
8. Ⓐ Ⓑ Ⓒ Ⓓ
9. Ⓐ Ⓑ Ⓒ Ⓓ
10. Ⓐ Ⓑ Ⓒ Ⓓ
11. Ⓐ Ⓑ Ⓒ Ⓓ
12. Ⓐ Ⓑ Ⓒ Ⓓ

Test Practice 2

1. Ⓐ Ⓑ Ⓒ Ⓓ
2. Ⓐ Ⓑ Ⓒ Ⓓ
3. Ⓐ Ⓑ Ⓒ Ⓓ
4. Ⓐ Ⓑ Ⓒ Ⓓ
5. Ⓐ Ⓑ Ⓒ Ⓓ
6. Ⓐ Ⓑ Ⓒ Ⓓ
7. Ⓐ Ⓑ Ⓒ Ⓓ
8. Ⓐ Ⓑ Ⓒ Ⓓ
9. Ⓐ Ⓑ Ⓒ Ⓓ
10. Ⓐ Ⓑ Ⓒ Ⓓ
11. Ⓐ Ⓑ Ⓒ Ⓓ
12. Ⓐ Ⓑ Ⓒ Ⓓ

Test Practice 3

1. Ⓐ Ⓑ Ⓒ Ⓓ
2. Ⓐ Ⓑ Ⓒ Ⓓ
3. Ⓐ Ⓑ Ⓒ Ⓓ
4. Ⓐ Ⓑ Ⓒ Ⓓ
5. Ⓐ Ⓑ Ⓒ Ⓓ
6. Ⓐ Ⓑ Ⓒ Ⓓ
7. Ⓐ Ⓑ Ⓒ Ⓓ
8. Ⓐ Ⓑ Ⓒ Ⓓ
9. Ⓐ Ⓑ Ⓒ Ⓓ
10. Ⓐ Ⓑ Ⓒ Ⓓ
11. Ⓐ Ⓑ Ⓒ Ⓓ
12. Ⓐ Ⓑ Ⓒ Ⓓ

Test Practice 4

1. Ⓐ Ⓑ Ⓒ Ⓓ
2. Ⓐ Ⓑ Ⓒ Ⓓ
3. Ⓐ Ⓑ Ⓒ Ⓓ
4. Ⓐ Ⓑ Ⓒ Ⓓ
5. Ⓐ Ⓑ Ⓒ Ⓓ
6. Ⓐ Ⓑ Ⓒ Ⓓ
7. Ⓐ Ⓑ Ⓒ Ⓓ
8. Ⓐ Ⓑ Ⓒ Ⓓ
9. Ⓐ Ⓑ Ⓒ Ⓓ
10. Ⓐ Ⓑ Ⓒ Ⓓ
11. Ⓐ Ⓑ Ⓒ Ⓓ
12. Ⓐ Ⓑ Ⓒ Ⓓ

Test Practice 5

1. Ⓐ Ⓑ Ⓒ Ⓓ
2. Ⓐ Ⓑ Ⓒ Ⓓ
3. Ⓐ Ⓑ Ⓒ Ⓓ
4. Ⓐ Ⓑ Ⓒ Ⓓ
5. Ⓐ Ⓑ Ⓒ Ⓓ
6. Ⓐ Ⓑ Ⓒ Ⓓ
7. Ⓐ Ⓑ Ⓒ Ⓓ
8. Ⓐ Ⓑ Ⓒ Ⓓ
9. Ⓐ Ⓑ Ⓒ Ⓓ
10. Ⓐ Ⓑ Ⓒ Ⓓ
11. Ⓐ Ⓑ Ⓒ Ⓓ
12. Ⓐ Ⓑ Ⓒ Ⓓ

Test Practice 6

1. Ⓐ Ⓑ Ⓒ Ⓓ
2. Ⓐ Ⓑ Ⓒ Ⓓ
3. Ⓐ Ⓑ Ⓒ Ⓓ
4. Ⓐ Ⓑ Ⓒ Ⓓ
5. Ⓐ Ⓑ Ⓒ Ⓓ
6. Ⓐ Ⓑ Ⓒ Ⓓ
7. Ⓐ Ⓑ Ⓒ Ⓓ
8. Ⓐ Ⓑ Ⓒ Ⓓ
9. Ⓐ Ⓑ Ⓒ Ⓓ
10. Ⓐ Ⓑ Ⓒ Ⓓ
11. Ⓐ Ⓑ Ⓒ Ⓓ
12. Ⓐ Ⓑ Ⓒ Ⓓ

Answer Key

Page 4
1. 1/5
2. 2/3
3. 7/8
4. 7/8
5. 1/6
6. 2/7
7. 1/4
8. 9/12 or 3/4
9. 3/10
10. 5/7
11. 1/8
12. 4/5
13. 5/12
14. 2/9
15. 8/11

Page 5
1. <
2. <
3. >
4. <
5. <
6. <
7. >
8. >
9. <
10. <
11. >
12. >

Page 6
1. 3/9
2. 9/11
3. 4/10
4. 5/6
5. 7/10
6. 6/8
7. 5/7
8. 7/9
9. 1/4
10. 3/6
11. 6/7
12. 2/3

Page 7
1. 1/3
2. 4/7
3. 4/8 or 1/2
4. 1/6
5. 1/4
6. 2/7
7. 5/6
8. 6/8 or 3/4
9. 5/7
10. 2/4 or 1/2
11. 2/3
12. 3/5
13. 1/2
14. 3/4

Page 8
1. 4/6 or 2/3
2. 3/10
3. 9/11
4. 7/8
5. 4/9
6. 8/11
7. 4/8 or 1/2
8. 1/7
9. 1/9
10. 3/4
11. 7/8
12. 6/7
13. 4/5
14. 3/10
15. 11/12
16. 2/3
17. 4/6 or 2/3
18. 1/5
19. 3/10
20. 6/12 or 1/2
21. 2/9
22. 5/12
23. 5/11
24. 1/3
25. 5/9
26. 9/12 or 3/4
27. 5/11
28. 1/11
29. 3/8
30. 2/5
31. 7/9
32. 4/8 or 1/2
33. 9/10
34. 1/11
35. 9/12 or 3/4
36. 4/6 or 2/3
37. 3/5
38. 7/12
39. 2/9
40. 5/10 or 1/2
41. 4/12 or 1/3
42. 8/10 or 4/5
43. 7/10
44. 6/8 or 3/4

Page 9
1. 5, 10, 15, 20, 25, 30, 35, 40, 45, 50
10, 20, 30, 40, 50, 60, 70, 80, 90, 100
common denominators: 10, 20, 30
2. 2, 4, 6, 8, 10, 12, 14, 16, 18, 20
10, 20, 30, 40, 50, 60, 70, 80, 90, 100
common denominators: 10, 20
3. 3, 6, 9, 12, 15, 18, 21, 24, 27, 30
9, 18, 27, 36, 45, 54, 63, 72, 81, 90
common denominators: 9, 18, 27
4. 3, 6, 9, 12, 15, 18, 21, 24, 27, 30
4, 8, 12, 16, 20, 24, 28, 32, 36, 40
common denominators: 12, 24
5. 5, 10, 15, 20, 25, 30, 35, 40, 45, 50
7, 14, 21, 28, 35, 42, 49, 56, 63, 70
common denominator: 35
6. 4, 8, 12, 16, 20, 24, 28, 32, 36, 40
6, 12, 18, 24, 30, 36, 42, 48, 54, 60
common denominators: 12, 24, 36
7. 6, 12, 18, 24, 30, 36, 42, 48, 54, 60
7, 14, 21, 28, 35, 42, 49, 56, 63, 70
common denominator: 42
8. 2, 4, 6, 8, 10, 12, 14, 16, 18, 20
8, 16, 24, 32, 40, 48, 56, 64, 72, 80
common denominators: 8, 16
9. 8, 16, 24, 32, 40, 48, 56, 64, 72, 80
9, 18, 27, 36, 45, 54, 63, 72, 81, 90
common denominator: 72

Page 10
1. 28; 7; 4; 7/28 + 12/28 = 19/28
2. 24; 4; 3; 20/24 – 15/24 = 5/24
3. 9; 3; —; 6/9 – 6/9 = 0
4. 70; 10; 7; 60/70 – 14/70 = 46/70 = 23/35
5. 30; 10; 3; 20/30 – 3/30 = 17/30
6. 90; 9; 10; 63/90 – 60/90 = 3/90 = 1/30
7. 72; 9; 8; 54/72 – 40/72 = 14/72 = 7/36
8. 63; 7; 9; 28/63 – 18/63 = 10/63
9. 21; 3; 7; 9/21 – 7/21 = 2/21
10. 15; 5; 3; 10/15 – 6/15 = 4/15
11. 10; —; 2; 3/10 + 6/10 = 9/10
12. 42; 7; 6; 35/42 –18/42 = 17/42
13. 12; 4; 3; 4/12 – 3/12 = 1/12

Page 11
1. 1/18
2. 11/14
3. 3/20
4. 9/10
5. 1/30
6. 7/22
7. 7/30
8. 11/30
9. 2/15
10. 3/10
11. 29/36
12. 1/3
13. 1/36

Hidden Message: Fractions come in all sizes.

Page 12
1. 19/30
2. 19/30
3. 3/70
4. 37/40
5. 1/6
6. 7/30
7. 13/24
8. 61/99
9. 51/72
10. 1/20
11. 0
12. 9/10
13. 4/15
14. 2/9
15. 11/28
16. <
17. >
18. <
19. >
20. >
21. <
22. <
23. <
24. <
25. <
26. <
27. <
28. >
29. <
30. <

Page 13
1. 1/4
2. 3/24, 2/16
3. 25/30, 10/12
4. 4/16, 3/12
5. 6/10, 15/25
6. 8/28, 6/21
7. 6/9, 8/12
8. 9/12, 18/24
9. 5/30, 3/18
10. 5/20, 2/8
11. 4/18, 6/27
12. 2/6, 3/9
13. 15/25, 6/10
14. 4/12, 6/18
15. 5/10, 50/100

Page 14
Note: The second answers shown for each problem may vary.
1. 1/4 or 2/8
2. 4/5 or 8/10
3. 3/4 or 6/8
4. 3/5 or 6/10
5. 1/4 or 2/8
6. 5/6 or 10/12
7. 3/8 or 6/16
8. 2/5 or 4/10
9. 1/2 or 3/6
10. 1/3 or 3/9
11. 1/5 or 2/10
12. 1/6 or 2/12
13. 2/3
14. 4/5
15. 3/4
16. 2/3
17. 1/6
18. 1/3
19. 2/5
20. 3/4
21. 1/2
22. 5/6
23. 1/5
24. 1/4

Page 15
1. 1/2
2. 1/3
3. 1/2
4. 3/5
5. 1/3
6. 1/8
7. 1/4
8. 1/3
9. 1/4
10. 1/9
11. 1/4
12. 2/9
13. 1/2
14. 1/6
15. 1/8

Page 16
1. 3 1/2
2. 8 1/3
3. 5 2/3
4. 3 1/3
5. 6 8/9
6. 6 1/8
7. 1 6/7
8. 1 1/5
9. 4 3/4
10. 3 7/9

Page 17
1. 3 1/4
2. 5 1/2
3. 7 3/5
4. 10 1/5
5. 2 1/4
6. 1 3/4
7. 8 1/2
8. 6 2/5
9. 4 2/3
10. 9 4/5
11. 1 5/6
12. 5 3/7
13. 7 5/6
14. 3 1/5

Page 18
1. 11 3/8
2. 5 2/3
3. 10 5/7
4. 4 1/2
5. 4
6. 5 1/10
7. 1 3/10
8. 12 2/7
9. 5 2/3
10. 14 1/2
11. 17 7/8
12. 1/5
13. 14 1/2
14. 11
15. 7 1/12
16. 2
17. 12 3/11
18. 18 11/12
19. 2
20. 5 3/4
21. 1 1/2
22. 7 1/3
23. 3 2/7
24. 14 7/10
25. 3/8

Page 19
1. 7
2. 10 3/4
3. 4
4. 7
5. 8
6. 11 2/9
7. 8 2/3
8. 8
9. 11
10. 2 3/4
11. 6 1/9
12. 5 1/2
13. 9 2/3
14. 11 2/5
15. 6 1/11
16. 10 5/12
17. 17 6/11
18. 13 1/4
19. 11
20. 18
21. 4 1/2
22. 8 3/5
23. 3 10/11
24. 4 5/6

Page 20
1. $.08
2. $10.82
3. $64.78
4. $.32
5. $3.66
6. $44.50
7. $.17
8. $9.49
9. $93.14
10. $3.81
11. $34.15
12. $51.72
13. $88/100
14. $3 12/100
15. $31 2/100
16. $55/100
17. $9 27/100
18. $81 89/100
19. $72/100
20. $5 32/100
21. $75 65/100
22. $2/100
23. $6 3/100
24. $23 57/100
25. $1.55
26. $.36
27. $1.47
28. $.24
29. $11.90
30. $11.00
31. $5.71
32. $9.81
33. $58.36
34. $100.96
35. $109.19
36. $48.54
37. $44.89
38. $39.96
39. $21.69

Page 21
1. 33/10, 7/3, 35/7
2. 17/4, 41/10, 12/5
3. 13/2, 10/9, 12/11
4. 20/5, 50/6, 8/3

Answer Key (cont.)

5. 1 2/5
6. 4 1/2
7. 3 2/5
8. 6 3/5
9. 4
10. 4 1/6
11. 3 6/7
12. 6 3/5
13. 6 4/11
14. 4 3/4
15. 5 1/2
16. 9 11/12
17. 51/8
18. 9/5
19. 87/10
20. 59/7
21. 97/12
22. 78/10
23. 19/12
24. 7/3
25. 5/4
26. 17/6
27. 68/11
28. 14/3
29. 27/12
30. 26/10

Page 22
1. 2 1/6
2. 1 1/3
3. 1 3/4
4. 1 3/5
5. 2 2/9
6. 8/9
7. 1 7/11
8. 1
9. 11/12
10. 1 7/11
11. 2
12. 1 7/8
13. 4/7
14. 5/6
15. 1 11/12
16. 2 1/6
17. 2 1/3
18. 1 4/5
19. 1 1/4
20. 3/5
21. 2 2/5
22. 1/3
23. 6
24. 3/5
25. 2 1/3
26. 1 2/9
27. 7/11
28. 5/9
29. 1 7/8
30. 2 3/4

Page 23
1. 6 1/3
2. 1 1/4
3. 9 3/11
4. 3
5. 15 1/6
6. 6 7/10
7. 1 17/33
8. 1 1/5
9. 16 1/3
10. 3 1/2
11. 11 1/2
12. 1/9
13. 2/3
14. 1/3
15. 1 5/9
16. 3
17. 4/5
18. 9/10
19. 1
20. 1 1/2
21. >
22. <
23. >
24. >
25. <
26. <
27. <
28. >
29. >
30. >
31. >
32. <
33. <
34. >
35. >

Page 24
1. 8 1/2
2. 4 1/4
3. 5 2/3
4. 7 3/4
5. 3 1/3
6. 9 1/4
7. 6 2/3
8. 2 1/2
9. 1 1/4
10–15. Make sure the fractions are marked correctly on the number line.
A. 1/3
B. 2 1/2
C. 4 1/3
D. 5 3/4
E. 1 1/2
F. 3 1/3
G. 5 1/2
H. 3 1/2
I. 2/3
J. 2 3/4
K. 5 1/4
L. 3 1/6

Page 25
1. 1/6
2. 1/60
3. 3/8
4. 6/35
5. 1/16
6. 27/40
7. 7/54
8. 3/16
9. 2/35
10. 2/9
11. 18/121
Hidden Message: It's as easy as one, two, three!

Page 26
1. 18/33
2. 28/99
3. 3/16
4. 2/9
5. 1/12
6. 15/49
7. 1/10
8. 1/12
9. 1/18
10. 1/24
11. 9/55
12. 28/55
13. 8/55
14. 7/20
15. 2/7
16. 3/10
17. 11/27
18. 2/55
19. 1/30
20. 3/8
21. 7/96
22. Jason gave away 22 gumballs. Jason has 2 gumballs left.
23. Penelope sold 24 yo-yos. She has 12 left.

Page 27
1. 1 1/11
2. 5 8/11
3. 1 7/18
4. 6/7
5. 18/25
6. 11/21
7. 1 4/5
8. 6/55
9. 5/12
10. 1 5/72
11. 7/10
12. 1 13/77
13. 2
14. 2/3
15. 2/3
16. 1 2/9
17. 3/5
18. 1 1/5
19. 1 1/10
20. 5/8

Page 28
1. 5/6
2. 6 2/3
3. 3 6/7
4. 27/32
5. 2/7
6. 11/45
7. 2/5
8. 15/16
9. 11/25
10. 3/4
11. 7/9
12. 1 7/15
13. 2 1/4
14. 1/4
15. 11/14
16. 1 25/96
17. 5/6
18. 3
19. 2 1/12
20. 1/3
21. 1/4 x 1/4 = 1/16   1/4 x 4/1 = 4/4 or 1
22. 1/2 x 1/2 = 1/4   1/2 x 2/1 = 2/2 or 1
23. 3/4 x 1/2 = 3/8   3/4 x 2/1 = 6/4 or 1 1/2
24. 5/6 x 1/9 = 5/54   5/6 x 9/1 = 15/2 or 7 1/2

Page 29
1. 8
2. 12
3. 4
4. 16
5. 15
6. 6
7. 35
8. 72
9. 40
10. 6
11. 56
12. 12
13. 9
14. 30
15. 20
16. 24
17. 8
18. 3

Page 30
1. 9
2. 10
3. 25
4. 24
5. 10
6. 8
7. 27
8. 54
9. 45
10. 23
11. 15
12. 25
13. 6
14. 17
15. 56
16. 12
17. 21
18. 15

Page 31
1. C
2. C
3. B
4. B
5. A
6. A
7. A
8. A
9. B
10. B
11. C
12. A

Page 32
1. B
2. A
3. B
4. A
5. B
6. C
7. C
8. C
9. A
10. C
11. A
12. B

Page 33
1. .67
2. .50
3. .75
4. .90
5. .40
6. .13
7. .10
8. .08
9. .80
10. .33
11. .11
12. .17
13. .83
14. .30
15. .20
16. .50
17. .43
18. .25
19. .60
20. .55

Page 34
2. 1 ÷ 3 = .33; 33%
3. 2 ÷ 3 = .67; 67%
4. 1 ÷ 4 = .25; 25%
5. 3 ÷ 4 = .75; 75%
6. 1 ÷ 5 = .20; 20%
7. 2 ÷ 5 = .40; 40%
8. 3 ÷ 5 = .60; 60%
9. 4 ÷ 5 = .80; 80%
10. 1 ÷ 6 = .17; 17%
11. 5 ÷ 6 = .83; 83%
12. 1 ÷ 8 = .13; 13%
13. 5 ÷ 8 = .63; 63%
14. 7 ÷ 8 = .88; 88%
15. 1 ÷ 9 = .11; 11%

Page 35
1. 4/5   .80   80%
2. 3/4   .75   75%
3. 1/6   .17   17%
4. 1/2   .50   50%
5. 3/7   .43   43%
6. 1/3   .33   33%
7. 7/10   .70   70%
8. 3/5   .60   60%
9. 4/7   .57   57%
10. 2/3   .67   67%
11. 1/8   .13   13%
12. 2/9   .22   22%
13. 1/4   .25   25%
14. 1/10   .10   10%
15. 4/9   .44   44%

Page 36
1. 2.5, <.50
2. 6, >.50
3. 3, >.50
4. 1.5, <.50
5. 3, <.50
6. 1.5, >.50
7. 5.5, >.50
8. 4.5, >.50
9. 2.5, <.50
10. 2, <.50
11. 3.5, >.50
12. 1, =.50
13. 2, >.50
14. 5.5, >.50
15. 4, >.50
16. 3.5, <.50

Page 37
1. 3.7, >.33
2. 1.7, <.33
3. 2.7, >.33

4. 3, = .33
5. 3, >.33
6. 3.7, <.33
7. 2.7, <.33
8. 2, =.25
9. 1.8, <.25
10. 5.8 <.25
11. 2.3, >.25
12. 2.3, <.25
13. 2.5, >.25
14. 2.8, >.25

Page 38
1. >
2. >
3. >
4. >
5. <
6. <
7. <
8. >
9. <
10. =
11. <
12. <
13. >
14. >
15. <
16. >
17. <
18. <
19. 1/2
20. 1/4
21. 1/3
22. 3/4
23. 1/5
24. 2/3
25. 2/5
26. 3/5
27. 4/5

Page 39
1. 8
2. 7
3. 2
4. 10
5. 22
6. 25
7. 14
8. 12

Page 40
1. B
2. C
3. A
4. A
5. B
6. B
7. A
8. C
9. B
10. A
11. B
12. C

Page 41
1. C
2. A
3. B
4. C
5. B
6. B
7. A
8. A
9. C
10. B
11. B
12. A

Page 42
1. B
2. B
3. A
4. A
5. C
6. A
7. B
8. C
9. C
10. A
11. A
12. B

Page 43
1. C
2. B
3. A
4. C
5. A
6. A
7. A
8. B
9. C
10. A
11. C
12. A

Page 44
1. B
2. A
3. C
4. B
5. A
6. B
7. A
8. B
9. A
10. C
11. C
12. C

Page 45
1. C
2. B
3. C
4. C
5. A
6. B
7. B
8. B
9. B
10. A
11. A
12. B